D1501219

The
Universe
in a
Handkerchief

The Universe in a Handkerchief

Lewis Carroll's
Mathematical Recreations,
Games, Puzzles,
and Word Plays

MARTIN GARDNER

COPERNICUS
AN IMPRINT OF SPRINGER-VERLAG

Published in the United States by Copernicus, an imprint of Springer-Verlag New York, Inc.

> Copernicus
> Springer-Verlag New York, Inc.
> 175 Fifth Avenue
> New York, NY 10010

The following materials are provided by the Harry Ransom Humanities Research Center, University of Texas at Austin, and are used by permission: Specific Gravities of Metals, &c. and Memoria Technica: for Numbers.

The following materials are provided by the Morris L. Parrish Collection, Department of Rare Books and Special Collections, Princeton University Libraries, and are used by permission: Word Links (pp. 89–91), Word Links (pp. 92–96), Doublets, Doublets Already Set, Preface to Glossary, Abbreviations, Glossary, Solutions of Doublets, Rules for Court Circular, Croquêt Castles, Castle-Croquêt, Lanrick (p. 138) Mischmasch, Syzygies, Circular Billiards, and Memoria Technica.

The following materials are provided by Fales Library, New York University, and are used by permission: Lanrick (p. 135), Lanrick (pp. 136–137), and Lanrick (pp. 139–140).

The pamphlets and other original works by Lewis Carroll are reproduced by kind courtesy of the Charles Lutwidge Dodgson estate.

Library of Congress Cataloging-in-Publication Data

Gardner, Martin, 1914–
    The universe in a handkerchief: Lewis Carroll's mathematical recreations, games, puzzles, and word plays/Martin Gardner.
      p. cm.
    Includes bibliographical references (p. 151–153) and index.
    ISBN 0-387-94673-X (hardcover: alk. paper)
    1. Mathematical recreations.   2. Literary recreations.
I. Carroll, Lewis, 1832–1898. II. Title.
QA95.G3325   1996
793.73—dc20   95-51303

Frontispiece: Lewis Carroll, a drawing by Harry Furniss, illustrator of Carroll's *Sylvie and Bruno*.

Manufactured in the United States of America.
Printed on acid-free paper.
Designed by Irmgard Lochner.

9 8 7 6 5 4 3 2 1

ISBN 0-387-94673-X   SPIN 10524577

To Clarkson N. Potter, friend and former publisher,
who in 1960 had the foresight to think it worthwhile
to annotate Lewis Carroll's *Alice* books.

Contents

Preface

I must confess that I did not become interested in Lewis Carroll until my undergraduate days at the University of Chicago. As a child, my greatest reading delights were the fantasies of L. Frank Baum. I tried hard to read the *Alice* books, but was put off by their abrupt transitions, the lack of a consistent story line, and the unpleasant characters in Alice's two dreams. And of course I missed all of Carroll's subtle jokes, word play, logic paradoxes, and philosophical implications. Unlike many Carrollians, I still believe that the *Alice* books should not be read by children, at least not by American children, until they are well into their teens.

When I reread the *Alice* books in my twenties, I was astounded by what I had missed. Two decades later, while writing a column on recreational mathematics for *Scientific American*, I discovered that Carroll not only shared my enthusiasm for play mathematics (puzzles, paradoxes, games, and so on), he also shared my hobby of conjuring. The more I learned about his life and opinions, the more I came to feel a spiritual kinship with him.

It occurred to me some 35 years ago that it was impossible for an American reader today, so far removed from Victorian England in both time and space, to appreciate fully the hundreds of hidden jokes in the *Alice* books without the aid of footnotes. I proposed the idea of an *Annotated Alice* to several publishers. They found the notion ridiculous. Scholarly notes on two simple children's books? What is there to say?

Clarkson Potter, then with Dial Press, was the first editor who did not think my proposal absurd. When he left Dial to form his own company, Clarkson Potter, Inc. (now a subdivision of Crown), he took my manuscript with him. *The Annotated Alice* was an instant success and has remained in print ever since. In 1990 I followed it with *More Annotated Alice*, with all new notes, with illustrations by Peter Newell instead of John Tenniel to distinguish the book's format from its predecessor.

Two books have been published about Carroll's mathematical and verbal play: *The Magic of Lewis Carroll*, by magician John Fisher, and *Lewis Carroll's Games and Puzzles*, by Edward Wakeling. Although there is overlap in what is offered in those two books and this one, I have organized the topics differently and included, as the other two books do not, the full texts of Carroll's privately published pamphlets and leaflets. I have also covered in detail the recreational aspects of Carroll's fiction, verse, letters, and magazine articles.

Literature about Carroll shows no signs of abating. Morton Cohen's long-awaited biography, issued in 1995 by Knopf, is packed with startling new revelations. One continues to be amazed by how much there is yet to learn about the life and writings of this shy, stammering teacher of mathematics, who for so long was regarded as little more than a scribbler of outlandish nonsense tales for children, an author too unimportant for scholars to take seriously.

1

Fiction and Verse

Charles Lutwidge Dodgson, who taught mathematics at Christ Church, Oxford, was a competent mathematician though not a great creative one. His original contributions were mainly in the recreational field. His strong sense of mathematical beauty became intertwined with a delight in play that found expression in a fondness for mathematical games, puzzles, logic paradoxes, magic tricks, riddles, and every variety of word play, especially puns, anagrams, and acrostic verse, published under the name Lewis Carroll.

Carroll's interest in card games and chess, as we all know, provided the background for his two immortal *Alice* books. Cards and chess pieces both have their kings and queens. In the first *Alice* book, the royalty and the Knave of Hearts, even the palace gardeners, are playing cards. Alice shouts at the court just before she awakens from her dream, "You're nothing but a pack of cards!" In the second *Alice* book, the kings, queens, and knights are chess pieces. The book's plot follows the moves of a whimsical, unorthodox chess game that culminates when Alice, who has the role of a white pawn, reaches the board's final rank to be crowned a queen.

Alice's Adventures in Wonderland swarms with word play, mostly obvious puns, although some are not so obvious. An example of puns so well concealed that they were long unrecognized are the three "littles" in the prefatory poem. They refer to the three Liddell sisters (Liddell rhymes with fiddle) who are in the boat with Carroll as they row up the Thames.

> All in the golden afternoon
>   Full leisurely we glide;
> For both our oars, with little skill,
>   By little arms are plied,
> While little hands make vain pretence
>   Our wanderings to guide.

The first *Alice* book also contains the Mad Hatter's notorious riddle about the raven and the writing desk. Carroll confessed that he introduced the riddle without having any answer in mind, though he later supplied one: "Because it can produce a few notes, tho they are *very* flat; and it is nevar put with the wrong end in front." Carroll deliberately misspelled "never" to make it "raven" backward. Scores of clever answers to the riddle have since been suggested by others.

Through the Looking-Glass is even richer in mathematical humor than the first *Alice* book. This is partly due to its pervasive chess themes, but also to the fact that Alice's journey into

[2]

the reversed world behind the mirror allowed Carroll to indulge in all sorts of bizarre reversals of space and time. Left and right symmetries and asymmetries abound. The Tweedle brothers, for instance, are mirror images of each other. The White Queen's memory works both forward and backward in time. She screams with pain *before* the pin of her brooch pricks her finger.

Mathematicians are always losing their way in endless labyrinths. The dozing Red King dreams about Alice, who is asleep and dreaming about the Red King. In both dreams, each dreams

The Tweedle brothers are what geometers call enantiomorphs—mirror reflections of each other. Here they are about to fight a battle. Note how John Tenniel, the illustrator, has carefully drawn them as mirror images.

of the other, forming a pair of infinite regresses. The book ends with Alice considering the "serious question" of which of them dreamed the other.

Linguistic play (which can be considered a branch of com-

Peter Newell's picture of the Red King as he dreams about Alice, while Alice in turn is dreaming about the sleeping Red King.

binatorics) and logical paradoxes pervade the second *Alice* book even more than the first. It swarms with puns and closes with a poem that is an acrostic on Alice's full name. The book also contains an unanswered conundrum; only this time Carroll knew the answer. The White Queen recites a riddle poem about a fish. The answer—an oyster—was not disclosed in print until it appeared anonymously in the magazine *Fun* (October 30, 1878).

In the two *Sylvie and Bruno* books, in which the real world and a fantasy realm are cleverly interwoven, Carroll's use of recreational mathematics, logic, and word play reaches still greater heights. The first volume's prefatory poem takes up where the terminal poem of *Through the Looking-Glass* leaves off. The three words that end its first stanza repeat, in reverse order, the three words that end the last lines of the earlier poem. Like the former poem, it too is an acrostic, an ingenious one. Not only do the first letters of the lines spell "Isa Bowman," one of Carroll's cherished child-friends, but the first three letters of each stanza are Isa, Bow, and Man.

The prefatory poem of *Sylvie and Bruno Concluded* is another unusual acrostic. The third letters of each line spell "Enid Stevens." The dedicatory poem of *The Nursery Alice* is still another acrostic, its lines' second letters spelling "Marie Van Der Gucht." Carroll wrote dozens of acrostic poems on children's names, which he sent to them in letters or inscribed in gift books. You'll find some of them gathered in a section on acrostics in the Modern Library edition of Carroll's writings.

Recreational mathematics is most explicit in the inventions of the Professor in the first *Sylvie and Bruno* book and of his counterpart, Mein Herr, a German professor in the sequel. In

A boat, beneath a sunny sky
Lingering onward dreamily
In an evening of July—

Children three that nestle near,
Eager eye and willing ear,
Pleased a simple tale to hear—

Long has paled that sunny sky;
Echoes fade and memories die;
Autumn frosts have slain July.

Still she haunts me, phantomwise,
Alice moving under skies
Never seen by waking eyes.

Children yet, the tale to hear,
Eager eye and willing ear,
Lovingly shall nestle near.

In a Wonderland they lie,
Dreaming as the days go by,
Dreaming as the summers die:

Ever drifting down the stream—
Lingering in the golden gleam—
Life, what is it but a dream?

Carroll's best-known acrostic poem, which he wrote as an epilogue to Through
the Looking-Glass. *The first letters of the lines spell the real Alice's name.*

the sequel's Chapter 7, Mein Herr explains to Lady Muriel
how a Möbius strip has only one side and one edge. He then
teaches her how to sew together two handkerchiefs to make a

Mein Herr shows Lady Muriel how to fold and sew a handkerchief to make a closed surface that has no outside or inside. The drawing is by Harry Furniss for Carroll's Sylvie and Bruno Concluded.

three-dimensional one-sided surface known to topologists to-day as a projective plane. (It is a close cousin of the better-known one-sided surface called a Klein bottle.) Mein Herr calls it Fortunatus's Purse because, having neither outside nor inside, it can be said to contain the entire universe. Carroll drew a sketch of the purse in a letter to the book's illustrator, Harry Furniss, who copied it exactly for his picture of the scene.

In the same chapter, Mein Herr describes a plan for running trains entirely by gravity. The track goes through a straight tunnel between two widely separated locations. Gravity pulls the train down to the tunnel's center, giving it sufficient momentum to continue up to the other end. Curiously, if friction and air resistance are ignored, the train will go from one end of the tunnel to the other in about 42 minutes regardless of the

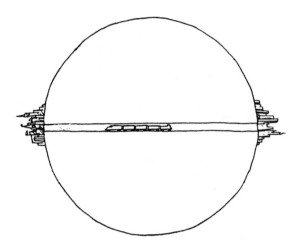

The gravity-operated train invented by the German professor in Carroll's Sylvie and Bruno Concluded. *From Martin Gardner's* Space Puzzles *(Simon and Schuster, 1971).*

tunnel's length. As we shall see, 42 had for Carroll some sort of special significance.

In Chapter 11 of *Sylvie and Bruno Concluded,* Mein Herr describes a map drawn on a scale of a mile to a mile:

> "It has never been spread out, yet," said Mein Herr: "the farmers objected; they said it would cover the whole country, and shut out the sunlight! So we now use the country itself, as its own map, and I assure you it does nearly as well."

Count Alfred Korzybski, founder of general semantics, liked to say "The map is not the territory." On the Professor's planet the two become identical.

Mein Herr goes on to describe a planet visited by a friend— a world so small that one can walk around it in twenty minutes:

> "There had been a great battle, just before his visit, which had ended rather oddly: the vanquished army ran away at

full speed, and in a very few minutes found themselves face-
to-face with the victorious army, who were marching home
again, and who were so frightened at finding themselves
between *two* armies, that they surrendered at once! Of course
that lost them the battle, though, as a matter of fact, they
had killed *all* the soldiers on the other side."

"Killed soldiers *ca'n't* run away," Bruno thoughtfully re-
marked.

"'Killed' is a technical word," replied Mein Herr. "In
the little planet I speak of, the bullets were made of soft
black stuff, which marked everything it touched. So, after a
battle, all you had to do was to count how many soldiers on
each side were 'killed'—that means 'marked on the *back*,' for
marks in *front* didn't count."

"Then you couldn't 'kill' any, unless they ran away?" I said.

"My scientific friend found out a better plan than
that. He pointed out that, if only the bullets were sent
the other way round the world, they would hit the enemy in
the *back*. After that, the *worst* marksmen were considered
the *best* soldiers; and *the very worst of all* always got First
Prize."

"And how did you decide which was *the very worst of all?*"

"Easily. The *best* possible shooting is, you know, to hit
what is exactly in *front* of you; so of course the *worst* possible
is to hit what is exactly *behind* you."

"They were strange people in that little planet!" I said.

"They were indeed! Perhaps their method of *government*
was the strangest of all. In *this* planet, I am told, a Nation
consists of a number of Subjects, and one King: but, in the
little planet I speak of, it consisted of a number of *Kings,*
and one *Subject!*"

Other inventions of Mein Herr deserve mention: a carriage with oval wheels that cause it to imitate the pitch and roll
of a ship, and boots with umbrellas attached to their tops to
guard against horizontal rain. Economists may disagree over

the best way to end a depression, but in *Sylvie and Bruno* (Chapter 21) the Professor has a simple solution:

> The Professor brightened up again. "The Emperor started the thing," he said. "He wanted to make everybody in Outland twice as rich as he was before—just to make the new Government popular. Only there wasn't nearly enough money in the Treasury to do it. So *I* suggested that he might do it by doubling the value of every coin and bank-note in Outland. It's the simplest thing possible. I wonder nobody ever thought of it before! And you never saw such universal joy. The shops are full from morning to night. Everybody's buying everything."

In Chapter 2 of *Sylvie and Bruno*, the Professor describes his portable plunge bath into which a bather plunges head first. It fits so closely around the body that one can take a complete bath with only half a gallon of water. The Mad Hatter's watch was curious in showing the days of the month, but the Professor's Outlandish Watch (Chapters 21 and 23) is far more outlandish. When a magic reversal peg is turned, time is reversed and all events run backward.

Carroll's great nonsense ballad, *The Hunting of the Snark,* bristles with word play, logic paradoxes, and mathematical nonsense. When the Butcher (in Fit 5) tries to convince the Beaver that 2 plus 1 is 3, he adopts a procedure that starts with 3 and ends with 3. It is not apparent, unless you write an algebraic expression for the operations, that the process must end with the same number you start with.

The algebraic expression is:

$$\frac{(x + 7 + 10)(1{,}000 - 8)}{992} - 17.$$

The Butcher, in The Hunting of the Snark, *is proving to the Beaver that 2 plus 1 is 3. His reasoning is circular because the procedure he is using always yields the same final value as the number he starts with. Henry Holiday, who illustrated Carroll's nonsense ballad, has filled the picture with objects as well as persons whose names start with B. Note also the kittens playing with the Butcher's yellow kid gloves, and the "income tax" lizard rifling the Butcher's pocket.*

This simplifies to:

$$\frac{(x+17)(992)}{992} - 17.$$

The 992's cancel each other, leaving

$$x + 17 - 17, \text{ or } x.$$

Thus x can take any value, and the expression's value will be the same.

Carroll was fond of children (provided they were attractive little girls), of Tuesdays, and of the number 42. When the Baker (Fit 1) comes aboard the ship, he leaves on the beach 42 carefully packed boxes with his name "painted clearly on each." According to the third Fit, stanza five, the Baker is in his early forties. It has been suggested that the Baker represents Carroll himself, who was 42 when he began writing the ballad. The 42 boxes are the 42 years he left behind when his imagination joined the ship's crew. A Rule 42 is cited in the book's Preface. In the first *Alice* book, during the farcical trial of the Knave of Hearts, the King invokes Rule 42. The number enters Carroll's writing in many other places, but no one knows just why.

The map used by the Snark hunters is the opposite of the mile-to-a-mile map in *Sylvie and Bruno Concluded*. It is totally blank. And the last of Henry Holiday's illustrations is a hidden picture puzzle. You have to look carefully to make out the Baker's horror-stricken face as the Boojum's paw drags him into oblivion.

"Novelty and Romancement," one of Carroll's few short works of fiction, is built around a play on words. A man sees a sign for "Roman Cement," which he mistakenly reads as "Romancement." The tale first appeared in Carroll's youthful

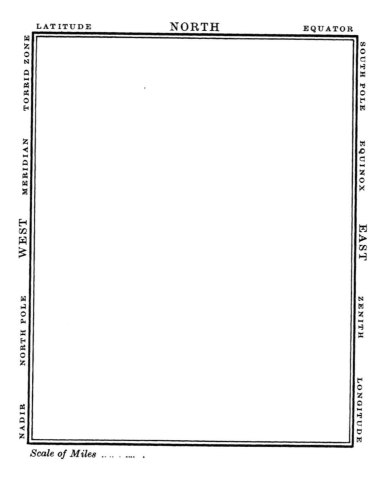

Scale of Miles ....

The map used by the crew hunting the Snark. It is totally free of errors.

periodical *The Train*, in 1856. You'll find it reprinted in the Modern Library edition of his works. It was published in Boston in 1925 as a booklet, with an Introduction by Randolph Edgar.

Carroll composed many verse charades. His earliest known original puzzle was a charade he published himself in his youthful journal *Mischmasch:*

Henry Holiday's illustration showing the vanishing of the Baker when he meets the Boojum in The Hunting of the Snark. *Do you see the Baker's agonized face concealed in the foliage as the Boojum's claw grasps his wrist? At upper left is the hand of the Bellman tingling his bell.*

A monument—men all agree—
Am I in all sincerity,
   Half cat, half hindrance made.
If head and tail removed should be,
Then most of all you strengthen me;
Replace my head, the stand you see
   On which my tail is laid.

The Lewis Carroll Handbook (Oxford University Press, 1962), by Sidney Williams, Falconer Madan, and Roger Green, provides the answer: Tablet.

A highly stylized verse form called a "double acrostic" was as popular in Carroll's day as charades. Each stanza provides a clue to a word. When the words are written down one under another their first letters form a word, and their last letters form another, related word. Horizontal words are called "cross-lights," and the two acrostic words are "uprights." Cross-lights may be of any length, but the two uprights must necessarily have the same number of letters. If the middle letters of the cross-lights spell a third word, the poem is called a triple acrostic.

The best known of Carroll's many double acrostics is mentioned in his diary (June 25, 1857). He said he was sitting alone in his room, listening to the music of a Christ Church ball, when he wrote a double acrostic for a "Miss Keyser." To this day its complete solution is not agreed upon. You'll find the poem analyzed in Chapter 6, on double acrostics, in my *Mathematical Magic Show*. Carroll included the poem in his 1883 collection of verse, *Rhyme? and Reason?*, but gave no solution.

THERE was an ancient City, stricken down
   With a strange frenzy, and for many a day
They paced from morn to eve the crowded town,
   And danced the night away.

I asked the cause: the aged man grew sad:
    They pointed to a building gray and tall,
And hoarsely answered "Step inside, my lad,
        And then you'll see it all."
                1

Yet what are all such gaieties to me
    Whose thoughts are full of indices and surds?

$$x^2 + 7x + 53$$
$$= \frac{11}{3}$$

                2

But something whispered "It will soon be done:
    Bands cannot always play, nor ladies smile:
Endure with patience the distasteful fun
        For just a little while!"
                3

A change came o'er my Vision—it was night:
    We clove a pathway through a frantic throng:
The steeds, wild-plunging, filled us with affright:
        The chariots whirled along.
                4

Within a marble hall a river ran—
    A living tide, half muslin and half cloth:
And here one mourned a broken wreath or fan,
        Yet swallowed down her wrath;
                5

And here one offered to a thirsty fair
    (His words half-drowned amid those thunders tuneful)
Some frozen viand (there were many there),
        A tooth-ache in each spoonful.
                6

There comes a happy pause, for human strength
    Will not endure to dance without cessation;
And every one must reach the point at length
        Of absolute prostration.
                7

At such a moment ladies learn to give,
　　To partners who would urge them overmuch,
A flat and yet decided negative—
　　　　Photographers love such.

8

There comes a welcome summons—hope revives,
　　And fading eyes grow bright, and pulses quicken:
Incessant pop the corks, and busy knives
　　　　Dispense the tongue and chicken.

9

Flushed with new life, the crowd flows back again:
　　And all is tangled talk and mazy motion—
Much like a waving field of golden grain,
　　　　Or a tempestuous ocean.

10

And thus they give the time, that Nature meant
　　For peaceful sleep and meditative snores,
To ceaseless din and mindless merriment
　　　　And waste of shoes and floors.

11

And One (we name him not) that flies the flowers,
　　That dreads the dances, and that shuns the salads,
They doom to pass in solitude the hours,
　　　　Writing acrostic-ballads.

12

How late it grows! The hour is surely past
　　That should have warned us with its double knock?
The twilight wanes, and morning comes at last—
　　　　"Oh, Uncle, what's o'clock?"

13

The Uncle gravely nods, and wisely winks.
　　It *may* mean much, but how is one to know?
He opes his mouth—yet out of it, methinks,
　　　　No words of wisdom flow.

The third stanza of this poem, for which the corresponding cross-light is probably "quadratic," has often been quoted as Carroll's whimsical portrayal of himself:

Yet what are all such gaieties to me
Whose thoughts are full of indices and surds?

$$x^2 + 7x + 53 = \frac{11}{3}$$

Unfortunately, no value of x will solve the quadratic equation. However, if the sign in front of 53 is changed from plus to minus, then the equation has two irrational solutions. Perhaps Carroll intended it to be −53, but after a printer got the sign wrong he allowed the mistake to remain. Or did he intend the equation to be nonsense?

Rhyme? and Reason? contains another double acrostic with cross-lights as much debated as those in the "Miss Keyser" puzzle. The solution is known to be "Ellen Terry," Carroll's longtime actress friend, because Carroll himself tells us that he wrote the poem after seeing Miss Terry play Ophelia in *Hamlet.* The cross-lights, however, which relate to Shakespeare's play, are far from clear.

EMPRESS of Art for thee I twine
   This wreath with all too slender skill.
Forgive my Muse each halting line,
   And for the deed accept the will!

O day of tears! Whence comes this spectre grim,
   Parting, like Death's cold river, souls that love?
Is not he bound to thee, as thou to him,
   By vows, unwhispered here, yet heard above?

And still it lives, that keen and heavenward flame.
  Lives in his eye, and trembles in his tone:
And these wild words of fury but proclaim
  A heart that beats for thee, for thee alone.

But all is lost: that mighty mind o'erthrown,
  Like sweet bells jangled, piteous sight to see!
'Doubt that the stars are fire,' so runs his moan,
  'Doubt Truth herself, but not my love for thee!'

A sadder vision yet: thine aged sire
  Shaming his hoary locks with treacherous wile!
And dost thou now doubt Truth to be a liar?
  And wilt thou die, that hast forgot to smile?

Nay, get thee hence! Leave all thy winsome ways
  And the faint fragrance of thy scattered flowers:
In holy silence wait the appointed days,
  And weep away the leaden-footed hours.

One of Carroll's most remarkable poems, if indeed he wrote it, was first published by Trevor Wakefield in his *Lewis Carroll Circular*, No. 2 (November 1974). The poem is quoted in a letter to *The Daily Express* (January 1, 1964) by a writer who tells of a privately printed book titled *Memoirs of Lady Ure*. Lady Ure, it seems, quoted the square poem as one that Carroll wrote for her brother. Wakefield says that no one has yet located a copy of Lady Ure's *Memoirs*, but whether this is still true I do not know. Here is the poem:

I often wondered when I cursed,
Often feared where I would be—
Wondered where she'd yield her love,
When I yield, so will she.
I would her will be pitied!
Cursed be love! She pitied me . . .

If you read this poem vertically—the first words of each line, then the second words, then the third, and so on—you get exactly the same poem as when you read the lines horizontally!

Some of the poems that a youthful Carroll contributed to an Oxford University periodical called *College Rhymes* were signed R. W. G. They are the fourth letters in each name of Charles Lutwidge Dodgson.

2

The Diaries

When Roger Green edited the two volumes of *The Diaries of Lewis Carroll* for Oxford University Press (1954), he left out many entries devoted to logic and mathematics, assuming they would hold little interest for readers. Happily, we will see the omitted entries when a new and complete edition of the *Diaries* is published under the editorship of Edward Wakeling. (Three volumes, covering the years 1855–1857, have come out as of this writing.) Meanwhile, we shall consider some of the paragraphs in the Green edition that are of special interest to recreational mathematicians and word play enthusiasts.

On December 19, 1898, Carroll wrote:

Sat up last night till 4 a.m., over a tempting problem, sent
me from New York, 'to find three equal rational-sided right-
angled triangles'. I found *two*, whose sides are 20, 21, 29;
12, 35, 37; but could not find three.

It turns out that there is an infinity of such triplets, but beyond
the three smallest in area the integral sides each have at least six
digits. The smallest solution has triangles of sides 40, 42, 58; 24,
70, 74; and 15, 112, 113. Their common area is 840. Had
Carroll doubled the sides of the two triangles he found, he
would have obtained the first two triangles in the triplet just
cited, from which it is easy to determine the third. A formula
for finding such triplets is given by Henry Dudeney in *Canter-
bury Puzzles* (London: Thomas Nelson, 1907, answer to Problem 107). See also "A Problem of Lewis Carroll's, and the
Rational Solution of a Diophantine Equation," by C. Tweedie,
in *Proceedings of the Edinburgh Mathematical Society,* Volume 24,
Session 1905–1906.

"I have worked out in the last few days," Carroll records
on May 27, 1894, "some curious problems on the plan of 'lying' dilemma. E.g., 'A says B lies; B says C lies; C says A and
B lie.'" The question is: Who lies and who tells the truth? One
must assume that A refers to B's statement, B to C's statement,
and C to the combined statements of A and B. The problem
was printed as an anonymous leaflet in 1894.

Only one answer does not lead to a logical contradiction:
A and C lie; B speaks the truth. The problem yields easily to
the propositional calculus by taking the word "says" as the
logical connective called equivalence. Without drawing on

symbolic logic one can simply list the eight possible combinations of lying and truth-telling for the three men, then explore each combination, eliminating those that lead to logical contradictions.

Carroll was fond of devising puzzles based on truth-tellers and liars. Many can be found in *Lewis Carroll's Symbolic Logic*, edited by the late William W. Bartley, III. Some of them are almost as bewildering as the clever truth/lie puzzles presented by mathematician and Carrollian Raymond Smullyan in his many puzzle books.

A problem in physics, hotly debated in Carroll's day, involves a monkey clinging to one end of a rope. The rope goes over a pulley with a weight on the other end that is equal to the monkey's weight. The monkey and the weight are at the

An illustration from Sam Loyd's Cyclopedia of Puzzles *(1914).*

same distance below the pulley. If the monkey climbs the rope, what happens to the weight? Carroll's diary on December 21, 1893, describes the problem, then adds: "It is very curious, the different views taken by good mathematicians. Price says the weight goes *up*, increasing velocity. Clifton (and Harcourt) that it goes *up*, at the same rate as the monkey, while Sampson says that it goes *down!*"

The correct answer, ignoring friction, is that regardless of how the monkey climbs—it may even let go of the rope and grab it again—the weight and monkey always stay at the same level. You can see this demonstrated by an exhibit in Chicago's Museum of Science and Industry.

On December 19, 1880, Carroll wrote: "The idea occurred to me that a game might be made of letters, to be moved about on a chess-board till they form words." There is no evidence that Carroll ever followed up this suggestion. In 1991 I tried to invent the kind of game Carroll had in mind. It is currently on sale as *Wordplay,* available from Kadon Enterprises, 1227 Lorene Drive, Pasadena, Maryland 21122.

On March 8, 1887, we find this entry: "Discovered a Rule for finding the day of the week for any given day of the month. There is less to remember than in any other Rule I have met with." Carroll published this rule as a short note in *Nature* (March 31, 1887). It later inspired the noted mathematician John Horton Conway, of Princeton University, to base on it a system for rapidly calculating in one's head the day of the week for any given date. Conway explains his method, with due credit to Carroll, in "Tomorrow is the Day After Doomsday," in the British periodical *Eureka,* No. 36, October 1973, pages 28–31.

Here is Carroll's note:

TO FIND THE DAY OF THE WEEK FOR ANY GIVEN DATE

Having hit upon the following method of mentally computing the day of the week for any given date, I send it you in the hope that it may interest some of your readers. I am not a rapid computer myself, and as I find my average time for doing any such question is about 20 seconds, I have little doubt that a rapid computer would not need 15.

Take the given date in 4 portions, viz. the number of centuries, the number of years over, the month, the day of the month.

Compute the following 4 items, adding each, when found, to the total of the previous items. When an item or total exceeds 7, divide by 7, and keep the remainder only.

The Century-Item.—For Old Style (which ended September 2, 1752) subtract from 18. For New Style (which began September 14) divide by 4, take overplus from 3, multiply remainder by 2.

The Year-Item.—Add together the number of dozens, the overplus, and the number of 4's in the overplus.

The Month-Item.—If it begins or ends with a vowel, subtract the number, denoting its place in the year, from 10. This, plus its number of days, gives the item for the following month. The item for January is "0"; for February or March (the 3rd month), "3"; for December (the 12th month), "12."

The Day-Item is the day of the month.

The total, thus reached, must be corrected, by deducting "1" (first adding 7, if the total be "0"), if the date be January or February in a Leap Year: remembering that every year, divisible by 4, is a Leap Year, excepting only the century-years, in New Style, when the number of centuries is *not* so divisible (*e.g.* 1800).

The final result gives the day of the week, "0" meaning Sunday, "1" Monday, and so on.

EXAMPLES

1783, *September* 18

17, divided by 4, leaves "1" over; 1 from 3 gives "2"; twice 2 is "4."

83 is 6 dozen and 11, giving 17; plus 2 gives 19, *i.e.* (dividing by 7) "5." Total 9, *i.e.* "2."

The item for August is "8 from 10," *i.e.* "2"; so, for September, it is "2 plus 3," *i.e.* "5." Total 7, *i.e.* "0," which goes out.

18 gives "4." Answer, "*Thursday.*"

1676, *February* 23

16 from 18 gives "2."

76 is 6 dozen and 4, giving 10; plus 1 gives 11, *i.e.* "4." Total "6."

The item for February is "3." Total 9, *i.e.* "2."

23 gives "2." Total "4."

Correction for Leap Year gives "3." Answer, "*Wednesday.*"

LEWIS CARROLL

Carroll was skilled in constructing anagrams on the names of people. Back in 1856 he recorded, in his youthful magazine *The Train,* two anagrams on his own first two names, Charles Lutwidge: "Edgar Cuthwellis" and "Edgar U.C. Westhill." He seriously considered using one of them as a pseudonym before settling on "Lewis Carroll."

A diary entry for November 25, 1868, reports that while "lying awake the other night," he thought of an anagram on William Ewart Gladstone: "Wilt tear down *all* images." Roger Green adds that Carroll later thought of a better one: "Wild agitator! Means well." In the same entry Carroll mentions hearing another anagram on Gladstone's name: "I, wise Mr. G, want to lead all!" This, he adds, can be answered by an anagram on Disraeli: "I lead, Sir!"

[26]

In his edition of the diaries, Green mentions in a note on Carroll's entry for October 17, 1870, that Rev. Edward Lee Hicks wrote in *his* diary, on October 22: "Heard this evening the last new joke of the author of *Alice in Wonderland*. He (Dodgson) knows a man whose feet are so large that he has to put on his trousers over his head." One can call this a joke about topology.

A note on Carroll's entry for August 29, 1897, describes another Irish Bull type of joke. Green quotes from a letter Carroll wrote to his sister Louisa: "Please analyse logically the following piece of reasoning: Little girl: 'I'm *so* glad I don't like asparagus . . . because, if I *did* like it, I should have to eat it—and I can't bear it!' It bothers me considerably."

The following entry appeared on June 20, 1892:

> Invented what I think is a *new* kind of riddle: 'A Russian had three sons. The first, named Rab, became a lawyer; the second, Ymra, became a soldier. The third became a sailor: what was his name?'

Rab is *bar* backward, and Ymra reverses to *army;* therefore the sailor's name is Yvan, a reversal of *navy*.

On February 4, 1894, Carroll mentions that he thought it would be a "pleasant variation in Backgammon to throw *three* dice [instead of two] and choose any two of the *three* numbers." He calls it "Thirdie Backgammon." On February 17 he changed the name to "Co-operative Backgammon," for which he invented a set of rules, published a few weeks later in *The Times* (March 6). Much earlier (on January 6, 1868) he mentions inventing a game he called "Blot-backgammon," but as far as I know its rules are not known.

Origami, or paper folding, may be viewed as a curious branch of geometry. On January 26, 1887, Carroll mentions that he folded a fishing boat (it had a seat at each end and a basket for fish in the middle) and a paper pistol for a little girl. On November 16, 1891, he writes of teaching the children of the Duchess of Albana how to "blot their names in creased paper," and how to fold paper pistols. The paper pistol is a

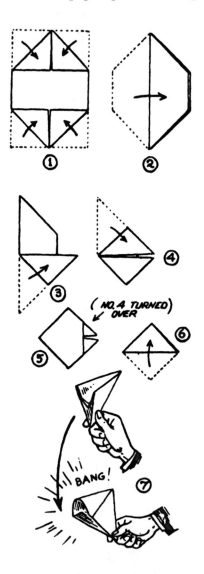

How to fold paper pistols. From Martin Gardner's Encyclopedia of Impromptu Magic *(Magic, Inc., 1978).*

sheet folded a certain way so that, when swung through the air, part of the fold pops out to make a sound like a pistol shot. He also showed the children "a machine which, by rapid spinning, turns the edging of a cup, etc., into a filmy solid."

Two folded paper hats, which Carroll surely knew how to make, were drawn by Tenniel for the second *Alice* book. The man dressed in white paper, in the railway carriage scene of Chapter 3, wears a paper hat; and the Carpenter, in the next chapter, wears a different kind of paper hat. Carpenters no longer fold such hats, but operators of printing presses, at least in the United States, like to make them from unprinted sheets to keep ink out of their hair.

Handkerchiefs and cloth napkins can be folded to make a variety of objects. In *The Story of Lewis Carroll* (London, J.M. Dent, 1899), Isa Bowman recalls seeing Carroll fold a handkerchief into the shape of a mouse and then skillfully make it seem to jump from his hand while he stroked its back. You'll find instructions for making this mouse in John Fisher's *The Magic of Lewis Carroll,* as well as in books on handkerchief magic.

In an entry for March 29, 1855, Carroll listed fourteen books he hoped to write. One uncompleted project was *Plain Facts for Circle-Squarers.* It would, he said, contain a proof that pi was between the limits of 3.14158 and 3.14160. Like so many mathematicians before and since, Carroll was plagued by cranks who believed they had found a way to square the circle with a compass and straightedge, an impossible task second only in notoriety to trisecting the angle. In his introduction to *A New Theory of Parallels,* Carroll recalls his correspondence with two such cranks, adding the following words, with which all mathematicians will surely agree:

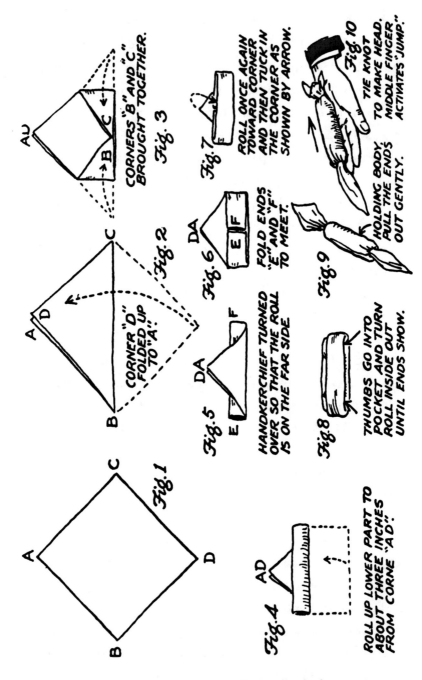

Fig. 1

A — C
B — D

Fig. 2

CORNER "D" FOLDED UP TO "A".

A, D
B — C

Fig. 3

CORNERS "B" AND "C" BROUGHT TOGETHER.

AD
B, C

Fig. 4

ROLL UP LOWER PART TO ABOUT THREE INCHES FROM CORNE "AD".

AD

Fig. 5

HANDKERCHIEF TURNED OVER SO THAT THE ROLL IS ON THE FAR SIDE

DA
E F

Fig. 6

FOLD ENDS "E" AND "F" TO MEET.

DA
E F

Fig. 7

ROLL ONCE AGAIN TOWARD CORNER AND THEN TUCK IN THE CORNER AS SHOWN BY ARROW.

Fig. 8

THUMBS GO INTO POCKET AND TURN ROLL INSIDE OUT UNTIL ENDS SHOW.

Fig. 9

HOLDING BODY, PULL THE ENDS OUT GENTLY.

Fig. 10

TIE KNOT TO MAKE HEAD. MIDDLE FINGER ACTIVATES "JUMP".

How Carroll made a jumping mouse with a handkerchief. From Martin Gardner's Encyclopedia of Impromptu Magic *(Magic, Inc., 1978).*

The first of these two misguided visionaries filled me with a great ambition to do a feat I have never heard of as accomplished by man, namely to convince a circle squarer of his error! The value my friend selected for π was 3.2: the enormous error tempted me with the idea that it could be easily demonstrated to *be* an error. More than a score of letters were interchanged before I became sadly convinced that I had no chance.

Carroll's system of mnemonics, based on an earlier system by Richard Gray, allows one to translate numbers into easily remembered words. To help recall the words, Carroll composed two-line rhymes in which the word was prominent, and the couplet described in some way the number it symbolized. The first reference to this system that appears in *The Diaries of Lewis Carroll* is on October 15, 1875: "Sat up till nearly 2 a.m. making a *Memoria Technica* for logarithms of primes up to 41. I can now calculate in a few minutes almost any logarithm without book." It has been said that Carroll planned to write a book titled *Logarithms by Lightning: A Mathematical Curiosity*, but I have been unable to verify this. Carroll's system enabled him to memorize logarithms up to seven decimal places.

On May 31, 1877, he wrote in his diary: "Spent a good part of the day and of the night, on revising my *Memoria Technica*." The following day he wrote: "The new *Memoria Technica* works beautifully. I made rhymes for the foundations of all the Colleges (except Univ.). At night I made lines giving pi to 71 decimal places."

On June 27, 1877, he said he had written an account of his system using an "electric pen," called a cyclostyle. This document is reproduced here along with a three-page pamphlet Carroll had printed a year later. These are followed by a cyclostyled leaflet, undated, giving couplets for memorizing the specific gravities of fifteen metals.

MEMORIA TECHNICA.
for Numbers.

1	2	3	4	5	6	7	8	9	0
b	d	t	f	l	s	p	h	n	z
c	w	j	qu	v	x	m	k	g	r

Each digit is represented by one or other of two consonants, according to the above table: vowels are then inserted *ad libitum* to form words, the significant consonants being always at the end of a line: the object of this is to give the important words the best chance of being, by means of the rhyme, remembered accurately.

The consonants have been chosen for the following reasons.

(1) *b, c,* first two consonants.

(2) *d* from "deux"; *w* from "two"

(3) *t* from "trois"; *j* was the last consonant left unappropriated.

(4) *f* from "four"; *qu* from "quatre".

(5) *l* = 50 ; *v* = 5.

(6) *s, x,* from "six."

(7) *p, m,* from "septem."

(8) *h* from "huit"; *k* from ὀκτώ.

(9) *n* from "nine"; *g* from its shape.

(0) *z, r,* from "zero."

They were also assigned in accordance, as far as possible, with the rule of giving to each digit one consonant in common use, and one rare one.

Since *y* is reckoned as a vowel, many whole words, (such as "ye", "you", "eye"), may be put in to make sense, without interfering with the significant letters.

Take as an example of this system the two dates of "Israelites leave Egypt — 1495," and "Israelites enter Canaan — 1455" :—

> "Shout again! We are free!"
> Says the loud voice of glee.
> "Nestle home like a dove,"
> Says the low voice of love.

Ch. Ch.
June 27/77

Memoria Technica.

My "Memoria Technica" is a modification of Gray's: but, whereas he used both consonants and vowels to represent digits, and had to content himself with a syllable of gibberish to represent the date or whatever other number was required, I use only consonants, and fill in with vowels "ad libitum", and thus can always manage to make a real word of whatever number has to be represented.

The principles, on which the necessary 20 consonants have been chosen, are as follows:--

[1] "b" and "c", the first two consonants in the Alphabet.
[2] "d" from "duo"; "w" from "two".
[3] "t" from "tres"; the other may wait awhile.

[4] "f" from "four"; "q" from "quatuor".

[5] "l" and "v", **because** "L" and "V" are the Roman symbols for "fifty" and "five".

[6] "s" and "x", from "six".

[7] "p" and "m", from "septem".

[8] "h" from "huit"; & "k" from the Greek "okto".

[9] "n" from "nine"; & "g", because it is so like a "9".

[0] "z" and "r", from "zero".

There is now one consonant still waiting for its digit, viz. "j"; and one digit waiting for its consonant, viz. "3": the conclusion is obvious.

The result may be tabulated thus:

1	2	3	4	5	6	7	8	9	0
b	d	t	f	l	s	p	h	n	z
c	w	j	q	v	x	m	k	g	r

When a word has been found, whose last consonants represent the number required, the best plan is to put it as the last word of a rhymed couplet,

so that , whatever other words in it
are forgotten, the rhyme will secure
the only really important word.

Now suppose you wish to remember
the date of the discovery of America,
which is "1492": the "1" may be left
out as obvious: all we need is "492".
Write it thus:--

4   9   .2

f   n   d
q   g   w

and try to find a word that contains
"f"or"q", "n"or"g", "d"or"w". A word
soon suggests itself---"found".

The poetical faculty must now be
brought into play, and the following
couplet will soon be evolved:--
"Columbus sailed the world around,
Until America was FOUND".

If possible, invent the couplets
for yourself: you will remember them
better than any made by others.

June, 1888.

Specific Gravities of Metals, &c.

[Water is taken as the unit. Translate into numbers the last four consonants of the couplet, and place a decimal point in the middle. e.g. Gold = 19·36]

Gold	Would you have enough Gold for your rents? Invest in the seven per cents.
Silver	With Silver the young soldier tip, And with a new sabre equip.
Copper	I bet you a copper that nook Is the place where the salmon broke hook.
Tin	It is merely a question of Tin, Where the wealthier suitor may win.
Lead	Leaden shot had been busy that day, Where many a dead rabbit lay.
Iron	Yes, Iron's the metal, old stoker, To make a superior poker!
Brass	Brass trumpet and brazen bassoon Will speedily mark you a tune.
Mercury	Quicksilver is quicker by far Than the liveliest live rabbits are.
Platinum	Of Platinum's little to spare: It is a commodity rare.
Lithium	From Lithium dread no fatigue, Though a lump you should carry a league.
Glass	Spun Glass will delight girl or boy: Nothing else makes so pretty a toy.
Deal	Yes, Deal is the timber, old mate, To make you a door or a gate.
Cork	This Cork to my shoulders I tie, And all marine terror defy.
Alcohol	For Alcohol cherish a dread; In excess it will injure your head.
Sea-water	Sea-water is held in repute An invalid's health to recruit.

The cyclostyle was originally a patented pen at the tip of which was a small wheel with a serrated edge. A page with writing was placed on a metal plate. When the writing was traced by the wheel it produced a stencil of tiny holes along the lines. The stencil was then placed on a blank sheet and an ink roller passed over it to make a copy.

Carroll later replaced his wheel pen with Thomas Edison's improved cyclostyle, called the "electric pen." A fine needle, battery operated, moved rapidly in and out of the pen to make the stencil holes. Thousands of copies could be made. The pen was widely sold here and abroad until it was replaced by type-writer-produced stencils and the mimeograph machine. (See "The Electric Pen," by Morton Cohen, in the *Illustrated London News,* Christmas 1976, pages 33–35.)

On February 26, 1858, Carroll wrote in his diary: "Invented another cipher, far better than the last." He cites four advantages: it is easily carried in the head, it uses a secret key word, no one can read it without knowing the key, and even if its English is known, the key word is impossible to discover. The last two advantages no longer hold for such ciphers, but in Carroll's time the art of cracking ciphers was in a primitive state. Green does not reprint the cipher, which apparently is in the diary, because it is "too long and complicated."

Ten years later (April 22, 1868) an entry reads: "Sitting up at night I invented a new cipher, which I think of calling *The Telegraph-Cipher.*" This was printed anonymously on two sides of an undated white card, presumably in 1888. An improved version titled "Alphabet-Cipher," printed on both sides of an undated white card, apparently was issued the same year, also anonymously. Carroll did not know at the time that he had reinvented a system familiar to cryptographers as the Vignère code.

Carroll's two cards are reprinted here. I will say no more about them because they are discussed at length in Francine Abeles' *The Mathematical Pamphlets of Charles Lutwidge Dodgson and Related Pieces,* published in 1994 by The Lewis Carroll Society of North America.

Carroll devoted considerable time to searching for new rules to test whether a large number is divisible by certain smaller numbers, especially primes. On June 3, 1884, he calls it an "inventive day" because he "concocted a new proportional rep-

THE TELEGRAPH-CIPHER.

DIRECTIONS FOR USE.

Cut this card in two along the line.

In order to send messages in this cipher, a key-word (or sentence) must be agreed on between the correspondents: this should be carried in the memory only.

To translate a message into cipher, write the key-word, letter for letter, over the message, repeating it as often as may be necessary: slide the message-alphabet along under the other, so as to bring the first letter of the message under the first letter of the key-word, and copy the letter that stands over 'a': then do the same with the second letter of the message and the second letter of the key-word, and so on.

Translate the cipher back into English by the same process.

[T. O.

For example, if the key-word be 'war,' and the message 'meet me at six,' we write it thus :—

```
⎰w a r w   a r   w a   r w a⎱
⎱m e e t   m e   a t   s i x⎰
 k w n d  'on  wh  z o d
```

The cipher sent, 'kwndonwhzod,' may be re-translated by the same process.

KEY-ALPHABET.

a b c d e f g h i j k l m n o p q r s t u v w x y z

a b c d e f g h i j k l m n o p q r s t u v w x y z a
MESSAGE-ALPHABET.

resentation scheme" and found rules for testing divisibility by 17 and 19. On September 27, 1897, he reports his discovery of a rule for dividing by 9 using only addition and subtraction, and another rule for division by 11. The next day he wrote that he had found even better rules for 9 and 11. On October 12, 1897, he completes his work on rules for dividing by 13, and on November 4 he improves the rule and generalizes it to divisors within 10 of any power of 10. Carroll's "Brief Method of Dividing a Given Number by 9 or 11" appeared in *Nature* on October 14, 1897.

On October 24, 1872, Carroll noted that he wrote out the rules for Arithmetical Croquet, a two-person game he had invented "a short time ago." He intended to put this in his never-finished book *Original Games and Puzzles*. Green reprints a manuscript copy dated April 22, 1899:

> 1. The first player names a number not greater than 8: the second does the same: the first then names a higher number, not advancing more than 8 beyond his last; and so on alternately—whoever names 100, which is 'winning peg', wins the game.
>
> 2. The numbers 10, 20, etc. are the 'hoops'. To 'take' a hoop, it is necessary to go, from a number below it, to one the same distance above it: e.g. to go from 17 to 23 would 'take' the hoop 20: but to go to any other number above 20 would 'miss it', in which case the player would have, in his next turn, to go back to a number below 20, in order to 'take' it properly. To miss a hoop twice loses the game.
>
> 3. It is also lawful to 'take' a hoop by playing *into* it, in one turn, and out of it, to the same distance above it in the next turn: e.g. to play from 17 to 20, and then from 20 to 23 in the next turn, would 'take' the hoop 20. A player 'in' a hoop may not play out of it with any other than the number so ordered.

THE
ALPHABET-CIPHER.

	A	B	C	D	E	F	G	H	I	J	K	L	M	N	O	P	Q	R	S	T	U	V	W	X	Y	Z	
A	a	b	c	d	e	f	g	h	i	j	k	l	m	n	o	p	q	r	s	t	u	v	w	x	y	z	**A**
B	b	c	d	e	f	g	h	i	j	k	l	m	n	o	p	q	r	s	t	u	v	w	x	y	z	a	**B**
C	c	d	e	f	g	h	i	j	k	l	m	n	o	p	q	r	s	t	u	v	w	x	y	z	a	b	**C**
D	d	e	f	g	h	i	j	k	l	m	n	o	p	q	r	s	t	u	v	w	x	y	z	a	b	c	**D**
E	e	f	g	h	i	j	k	l	m	n	o	p	q	r	s	t	u	v	w	x	y	z	a	b	c	d	**E**
F	f	g	h	i	j	k	l	m	n	o	p	q	r	s	t	u	v	w	x	y	z	a	b	c	d	e	**F**
G	g	h	i	j	k	l	m	n	o	p	q	r	s	t	u	v	w	x	y	z	a	b	c	d	e	f	**G**
H	h	i	j	k	l	m	n	o	p	q	r	s	t	u	v	w	x	y	z	a	b	c	d	e	f	g	**H**
I	i	j	k	l	m	n	o	p	q	r	s	t	u	v	w	x	y	z	a	b	c	d	e	f	g	h	**I**
J	j	k	l	m	n	o	p	q	r	s	t	u	v	w	x	y	z	a	b	c	d	e	f	g	h	i	**J**
K	k	l	m	n	o	p	q	r	s	t	u	v	w	x	y	z	a	b	c	d	e	f	g	h	i	j	**K**
L	l	m	n	o	p	q	r	s	t	u	v	w	x	y	z	a	b	c	d	e	f	g	h	i	j	k	**L**
M	m	n	o	p	q	r	s	t	u	v	w	x	y	z	a	b	c	d	e	f	g	h	i	j	k	l	**M**
N	n	o	p	q	r	s	t	u	v	w	x	y	z	a	b	c	d	e	f	g	h	i	j	k	l	m	**N**
O	o	p	q	r	s	t	u	v	w	x	y	z	a	b	c	d	e	f	g	h	i	j	k	l	m	n	**O**
P	p	q	r	s	t	u	v	w	x	y	z	a	b	c	d	e	f	g	h	i	j	k	l	m	n	o	**P**
Q	q	r	s	t	u	v	w	x	y	z	a	b	c	d	e	f	g	h	i	j	k	l	m	n	o	p	**Q**
R	r	s	t	u	v	w	x	y	z	a	b	c	d	e	f	g	h	i	j	k	l	m	n	o	p	q	**R**
S	s	t	u	v	w	x	y	z	a	b	c	d	e	f	g	h	i	j	k	l	m	n	o	p	q	r	**S**
T	t	u	v	w	x	y	z	a	b	c	d	e	f	g	h	i	j	k	l	m	n	o	p	q	r	s	**T**
U	u	v	w	x	y	z	a	b	c	d	e	f	g	h	i	j	k	l	m	n	o	p	q	r	s	t	**U**
V	v	w	x	y	z	a	b	c	d	e	f	g	h	i	j	k	l	m	n	o	p	q	r	s	t	u	**V**
W	w	x	y	z	a	b	c	d	e	f	g	h	i	j	k	l	m	n	o	p	q	r	s	t	u	v	**W**
X	x	y	z	a	b	c	d	e	f	g	h	i	j	k	l	m	n	o	p	q	r	s	t	u	v	w	**X**
Y	y	z	a	b	c	d	e	f	g	h	i	j	k	l	m	n	o	p	q	r	s	t	u	v	w	x	**Y**
Z	z	a	b	c	d	e	f	g	h	i	j	k	l	m	n	o	p	q	r	s	t	u	v	w	x	y	**Z**
	A	B	C	D	E	F	G	H	I	J	K	L	M	N	O	P	Q	R	S	T	U	V	W	X	Y	Z	

An explanation of the method of using the above table for sending Messages will be found on the other side.

EXPLANATION.

Each column of this table forms a dictionary of symbols representing the alphabet: thus, in the A column, the symbol is the same as the letter represented; in the B column, A is represented by B, B by C, and so on.

To use the table, some word or sentence should be agreed on by two correspondents. This may be called the "key-word," or "key-sentence," and should be carried in the memory only.

In sending a message, write the key-word over it, letter for letter, repeating it as often as may be necessary: the letters of the key-word will indicate which column is to be used in translating each letter of the message, the symbols for which should be written underneath: then copy out the symbols only, and destroy the first paper. It will now be impossible for any one, ignorant of the key-word, to decipher the message, even with the help of the table.

For example, let the key-word be *vigilance*, and the message "meet me on Tuesday evening at seven," the first paper will read as follows:—

```
v i g i l a n c e v i g i l a n c e v i g i l a n c e v i
m e e t m e o n t u e s d a y e v e n i n g a t s e v e n
h m k b x e b p x p m y l l y r x i i q t o l t f g z z v
```

the second will contain only "h m k b x e b p x p m y l l y r x i i q t o l t f g z z v."

The receiver of the message can, by the same process, retranslate it into English.

N.B. If this table be lost, it can easily be written out from memory, by observing that the first symbol in each column is the same as the letter naming the column, and that they are continued downwards in alphabetical order. Of course it would only be necessary to write out the particular columns required by the key-word: such a paper, however, should not be preserved, as it would afford means for discovering the key-word.

4. Whatever step one player takes, bars the other from taking an equal step, or the difference between it and 9: e.g. if one player advances 2, the other may not advance 2 or 7. But a player has no 'barring' power when playing *into* a hoop, or when playing from any number between 90 and 100, unless the other player is also at such a number.

5. The 'winning-peg', like the 'hoops', may be 'missed' once, but to miss it twice loses the game.

6. When one player is 'in' a hoop, the other can keep him in, by playing the number he needs for coming out, so as to bar him from using it. He can also do it by playing the difference between this and 9. And he may thus go on playing the 2 barring numbers alternately: but he may not play either twice running: e.g. if one player has gone from 17 to 20, the other can keep him in by playing 3, 6, 3, 6, etc.

In *The Lewis Carroll Picture Book* Ella Monier-Williams, a former child-friend and photographic subject, recalls a pleasant walk with Carroll when they played arithmetical croquet in their heads. "How it was done I cannot recollect, but his clever original brain planned it out by some system of mathematical calculation."

In 1994, the first two volumes of Carroll's unexpurgated diary were published in England. In the second volume, for the year 1856, I found the following entries, all omitted by Roger Green in his earlier edition.

On February 5 Carroll wrote:

Varied the lesson at the school with a story, introducing a number of sums to be worked out. I also worked for them the puzzle of writing the answer to an addition sum, when only one of the five rows have been written: this, and the trick of counting alternately up to 100, neither putting on more than 10 to the number last named, astonished them not a little.

The addition trick, then well known, starts with a magician writing down an arbitrary number of, say, five digits. Someone else puts any five-digit number beneath it. The magician then writes a third number, a spectator writes a fourth, and the magician adds a fifth. He is now instantly able to record the sum of the five five-digit numbers.

The secret is that for his third and fifth numbers the magician writes the complement of 9 for each digit of the number directly above. The final sum is obtained from the first number by taking 2 from its last digit and placing 2 at the front of the number.

For example:

```
  21879
  42351
  57648
  94366
  05633
 ------
 221877
```

The game of counting to 100 is a nim-like contest, a precursor of Carroll's Arithmetical Croquet, in which two players alternately call numbers from 1 through 10. A running total is kept. The winner is the player who reaches the sum of 100. The first player can always win by naming 1, and thereafter calling numbers that bring the partial sum to the following key numbers: 12, 23, 34, 45, 56, 67, 78, 89. The series is easy to recall. If a player not privy to the strategy starts with 1, the second player can of course win by playing to any of the key numbers, and then following the sequence until he reaches 100.

On February 8 Carroll said he entertained his class with "the 9 trick of striking out a figure after subtracting a number from its reverse." When a number is reversed and the smaller taken from the larger, digits in the remainder always add to a multiple of 9. With his back turned, the magician requests that any nonzero digit be crossed out of the remainder, and the digits left be read aloud in any order. In his head the performer adds the digits called out, casting out nines as the digits are called, and then subtracts this final digit from 9 to obtain the deleted digit.

The trick works even if the digits of the original number are scrambled. For example, I have just copied the following number from a dollar bill: 17240184. A scramble of the digits yields 47810412. Taking the smaller number from the larger results in 30570228, a multiple of 9. Adding the digits and casting out nines as you go along ends with a "digital root" of 9. Assume the number crossed out is 8. The digital root of the remaining digits will be 1. Taking 1 from 9 leaves 8.

The following entry is dated February 29:

> I have been trying for the last two days to solve a problem in chances, given me by Pember, which is said to have raised much discussion in the college. It is an exceedingly complicated question, and I have not yet got near a solution.
>
> Problem in the game of "Sympathy." The game is this: two players lay out two separate packs in heaps of 3, (and one card over in each pack), turning each top card face upwards, so as to have 18 faces on each side. Those which correspond are paired off together, and the cards under them turned face up: (the simplest way would be, to lay all face up originally).
>
> Required: the chance of the whole pack being paired off in this way.

Calling this a "complicated question" is an understatement. I have no idea how to go about solving it.

On April 8 Carroll wrote:

A letter appeared in *The Times,* from "Jellinger Symons," denying the rotation of the moon. I sent an answer, a sort of practical illustration of the necessity of its rotation. In considering the subject, I noticed for the first time the fact that though it only goes 13 times round the earth in the course of the year, it makes 14 revolutions round its own axis, the extra one being due to its motion round the sun.

To this entry Edward Wakeling, the new edition's editor, adds a note:

A letter under the title "The Moon Has No Rotary Motion" appeared in *The Times* on 8 April signed by Jelinger Symons, Her Majesty's Inspector of Schools. The letter begins: "May I request the favour of a small space in your columns to inquire the grounds upon which almost all school astronomy books assert that the moon rotates on her axis?" He goes on to say that since only one side of the moon's surface is ever visible from the earth, it does in fact not rotate on its axis. The following day *The Times* reported a vast postbag of replies and printed seven letters as a representative sample of the responses. Dodgson's reply did not appear. Jelinger Symons wrote again on 14 April still denying the rotation of the moon and Dodgson penned another reply. On 15 April, a letter was printed concentrating on the geometrical principles involved in the problem, but since it was signed by E.B.D. this is unlikely to be by Dodgson even accounting for a transcription error with his initials. The controversy rolled on throughout the month with further letters from Symons and other correspondents. Dodgson does not appear to have written any further letters, probably sensing that with two replies already rejected there is little chance of getting another contribution accepted.

The question of whether the moon rotates often generates hilarious parlor arguments. It resembles the old question of whether a hunter, who walks around a squirrel who keeps turning to face the hunter, has walked around the squirrel. As William James made clear when he discussed this question in his book *Pragmatism,* it all depends on what you mean by "around."

The moon problem is a similar vacuous debate over what is meant by "rotate." To an observer on the earth the moon does not rotate. To an observer on a star outside the solar system, the moon rotates once for each revolution around the earth.

Ten years after the controversy in England, the same arguments broke out in America. It is hard to believe, but for three years the question was debated in letters published in *Scientific American.* Not until 1868 did the editors announce that no more letters on the topic would be printed, but that a new magazine called *The Wheel* would be devoted entirely to this "great question." At least one issue appeared. I would love to own a copy, but it must be extremely rare. For an account of all this, see Chapter 16 of my *Mathematical Circus* (Mathematical Association of America, updated edition, 1992).

I have no doubt that more entries relating to recreational mathematics will turn up in later volumes of Wakeling's unexpurgated *Diaries.*

3

Letters

Lewis Carroll wrote thousands of letters, mostly to young women, that bristle with puns, riddles, acrostic verse, charades, anagrams, word play, and occasional mathematical puzzles. Sometimes the very form of writing was a puzzle. He sent letters that could be read only by holding them up to a mirror or turning the page over and looking through it at a strong light. Some were rebuses, with little pictures replacing words. Some had the words in proper order, but each word spelled backward. Others had to be read backward entirely, letter by letter. One was written in spiral form. Another had scraggly handwriting which Carroll pretended was caused by his hand shaking with fear over writing to the young lady. Occasionally he would compose a letter in one of his cipher

A spiral letter written by Lewis Carroll in 1878 to his child-friend Agnes Hull.

systems, providing the key word for decoding it. One surviving letter is in such tiny script that a magnifying glass is needed to read it.

Here is a sampling from the scores of anagrams in Carroll's letters:

In a letter to Mabel Scott he anagrams AMIABLEST? to 'TIS MABEL!, and WHERE MABEL? to WE BLAME HER.

He asks Mary Newby to rearrange the letters of NOR DO
WE to make one word. Answer: ONE WORD. (The letters
also spell NEW DOOR.)

To Maud Standen he sent a short poem containing twenty-
four single-word anagrams to be formed from two or three
adjacent words:

> As to the war, try elm. I tried.
> The wig cast in, I went to ride.
> "Ring? Yes." We rang. "Let's rap."
> We don't.
> "O shew her wit!" As yet she won't.
> Saw eel in Rome. Dry one; he's wet.
> I am dry. O forge! Th' rogue! Why a net?

To this day scholars are not agreed on all twenty-four words.

In the same letter to Maud he asked her to scramble the
letters ABCDEFGI to make a hyphenated word "as good as
summer-house." Answer: BIG-FACED.

In a letter to Enid Stevens he asked for single-word anagrams of DRY ONE, HE'S WET, and SCALE IT. Answers:
YONDER, SEWETH, and ELASTIC.

An anagram on the name of Edward Vaughan Kenealy, a
famous London barrister, was given in a letter to Francis Paget:
"Ah! We dread an ugly knave!" As he often did, Carroll says he
constructed it one night after going to bed.

Here is a sampling of riddles from Carroll letters:

"Why is Agnes like a thermometer?" Because she won't
rise when it's cold. (Agnes Hull hated to get up on chilly morn-
ings.)

Carroll asks Gertrude Chataway, "Why is a pig that has
lost its tail like a little girl on the seashore?" Because it says, "I
should like another tale, please!"

To the question "Why does Agnes know so much about

insects?" Carroll gave the convoluted answer, "Because *she* is so deep in entomology." The French word for "she" is "elle," and the letter L is deep (the seventh letter) inside the word "entomology"! "It couldn't be well deeper," Carroll added, "unless it happened to be in a deeper well."

How does a doll know that a hand which came off was her right hand? Because the other hand was left. (I failed to note in which letter I came across this.)

From myriad instances of word and number play in Carroll's letters, here are a few:

Make sense by punctuating "It was and I said not all," he asks Mary Newby. Answer: "It was *and*, I said, not *all*."

Carroll preferred to have little girls visit him alone. Inviting Dorothy Poole to come see him, he asked her not to be alarmed if the number of guests present would be .99999.... The infinite decimal fraction "*looks* alarming, I grant; but circulating decimals lose much of their grandeur when reduced to vulgar fractions!"

Carroll informs Mary Macdonald, age 21, that he was twice her age the previous year. When was he three times her age?

Edward Wakeling, in *Lewis Carroll's Games and Puzzles* (1992), includes a number of puzzles from Carroll's surviving letters to child-friends. Enid Stevens was asked to solve this problem:

> Three men, A, B, and C, are to run a race of a quarter-of-a-mile. Whenever A runs against B, he loses 10 yards in every 100; whenever B runs against C, he *gains* 10 yards in every 100. How should they be handicapped? ("Handicapping" means that the inferior runners are allowed a *start:* and the amount is so calculated that, if all were to run at their previous rates, it would be a *dead heat:* i.e. they would all get to the winning post at the same moment.)

Morton Cohen, commenting on this in his *The Letters of Lewis*

Carroll (Volume 2, page 1119), says that the problem is too ambiguously stated to have a precise answer.

Carroll was fond of what puzzlists call "river-crossing problems," and several turn up in his letters. His favorite was the old problem involving a fox, a goose, and a bag of corn. One

A wolf does not eat cabbage, so the crossing can start with the goat.

The man leaves the goat and returns, puts the cabbage in the boat, and takes it across. On the other bank, he leaves the cabbage but takes the goat.

He leaves the goat on the first bank and takes the wolf across. He leaves the cabbage with the wolf and rows back alone.

He takes the goat across.

The solution to Carroll's problem of the fox, goose, and bag of corn, here presented with a fox, goat, and cabbage. The illustration is from The Moscow Puzzles, *by Boris Kordemski (Scribner's, 1972).*

letter, to Winifred Hawke, contains Carroll's rules for playing a word game he invented. It involved drawing counters with vowels from one bag, counters with consonants from another, and using them to form words.

Still another letter, included in Wakeling's book, gave Helen Fielden a classic geometrical puzzle about a square window three feet on each side. The task is to alter the window to a square with half the area as before, yet remaining three feet high and three feet wide. The new window is a square tilted so the ends of its two diagonals touch the midpoints of the four sides of the original window. Because each diagonal is three feet, the window remains three feet high and three feet wide.

A pencil-drawing puzzle that Carroll loved to show to children, or send to them in letters, involves three squares interlaced like this:

The task is to draw the squares without taking the pencil off the paper, without retracing a line, or having any line cross another, and to return to the starting spot. In other words, the line must be topologically equivalent to a closed curve that does not self-intersect. The pattern, incidentally, is topologically the same as the three intersecting circles proposed by John Venn for diagramming problems in class-inclusion logic, and it is closely related to a diagram Carroll invented for solving syllogisms, as we shall see in the next chapter. Today such tracing tasks are regarded as problems in graph theory. The interlaced squares puzzle has several solutions, not difficult to discover.

Henry Ernest Dudeney, the famous English puzzle maker, included the interlaced squares in "Some Much Discussed Puzzles," an article in *Strand Magazine* (May 1908). He mentions that Carroll is often credited with inventing the puzzle, but he says that he found it in a "little book published in 1885." (What book, one wonders?) John Cook Wilson, Carroll's antagonist over a logic paradox we will discuss in the next chapter, wrote a book titled *On the Tracing of Geometrical Figures* (1905). I do not know if it includes the interlocked squares.

Wakeling's book published for the first time a series of puzzles in recently discovered letters from Carroll to his former mathematics teacher at Oxford, Professor Bartholomew Price. One involves a clock's reflection in a mirror, another asks for the number of different ways a cube can be colored with six colors, one on each face. The answer is 30. This set of thirty cubes has many unusual properties that have been explored by John Horton Conway and discussed in Chapter 6 of my *Fractal Music, Hypercards, and More.*

Two other problems in Carroll's letters to Price appear in Wakeling's book. One is a logic problem about six persons, with rules governing how they may go out of a house or stay inside. The other is a famous number problem, not original with Carroll, concerning four men, a monkey, and a pile of nuts. The problem has many variations. One involving five men, a monkey, and a pile of coconuts was the basis of a short story by Ben Ames Williams. I analyzed it in Chapter 9 of *The Second Book of Mathematical Puzzles and Diversions from Scientific American.*

Morton Cohen, in the first volume of *The Letters of Lewis Carroll* (page 577), prints a letter in which Carroll poses for Wilton Rix (one of his few letters to a boy) the following algebraic fallacy:

Understanding you to be a distinguished algebraist (i.e. distinguished from other algebraists by different face, different height, etc.) I beg to submit to you a difficulty which distresses me much.

If x and y are each equal to "1," it is plain that $2 \times (x^2 - y^2) = 0$ and also that $5 \times (x - y) = 0$. Hence $2 \times (x^2 - y^2) = 5 \times (x - y)$.

Now divide each side of this equation by $(x - y)$.

Then $2 \times (x + y) = 5$.

But $(x + y) = (1 + 1)$, i.e. $= 2$.

So that $2 \times 2 = 5$.

Ever since this painful fact has been forced upon me, I have not slept more than 8 hours a night, and have not been able to eat more than 3 meals a day.

I trust you will pity me and will kindly explain the difficulty to

Your obliged,
Lewis Carroll.

The error arises from the fact that x minus y equals zero, and one of the steps allows a division by zero, which the rules of algebra do not permit.

In a letter to Annie Rogers, Carroll opened with an acrostic poem based on ABCDE:

I send you
A picture, which I hope will
B one that you will like to
C. If your Mamma should
D sire one like it, I could
E sily get her one.

Writing to Violet Butler, Carroll lettered her name five times, placing the names so that "Olive," the name of her eldest sister, could be read vertically:

VIOLET
VIOLET
VIOLET
VIOLET
VIOLET

Ink cannot contain *ink,* he wrote to Gertrude Chataway, but *wink* does indeed contain *ink.* The letter ends with "I send you 10,000,000 kisses." Six days later, in another letter to Gertrude, he reduces the number to 4¾ kisses.

Morton Cohen's footnote on page 114 of the first volume of *The Letters of Lewis Carroll* reports on a puzzle Carroll sent to Margaret Cunnynghame. It consists of two sets of words, each lettered on a cardboard rectangle. The task is to arrange the words of each set to make an English sentence. The two solutions are: "The last time I offered Maggie some pudding she saucily replied that she didn't care twopence for it," and "If the man plays Mozart all night he will have hard work."

In one of his best-known letters to Maggie, Carroll sketched a caricature of his face, his right hand covering all of the face except the eyes. Although this lengthy letter appears to be in prose, and is usually read as such, it actually is entirely in meter and rhyme. Even the postscript is in verse:

My love to yourself—to your Mother
My kindest regards—to your small,
Fat, impertinent, ignorant brother
My hatred. I think that is all.

The postscript has been quoted by Carroll scholars who did not realize it was verse. The entire letter may be found in *The Letters of Lewis Carroll* (Volume 1, pages 112–113), and in Collingwood's *The Life and Letters of Lewis Carroll.*

In a letter included in the first volume of *The Letters of Lewis Carroll* (page 605), Edith Rix is asked what she can conclude from the following premisses:

No healthy Englishmen are hermits:
All strong hermits are healthy:
All healthy Englishmen are strong.

I am embarrassed to say that in my solution to this problem, which Professor Cohen published as a footnote, I blundered. Let's assign the following letters to the five terms of the problem:

A. Healthy Englishmen.
B. Hermits.
C. Strong hermits.
D. Strong people.
E. Healthy people.

The three premisses can now be written:

1. No A is B.
2. All C is E.
3. All A is D.

To these statements we can add:

4. All C is B.
5. All A is E.
6. All C is D.

I analyzed the problem by shading regions on a Venn diagram for five terms. All regions where the set of healthy Englishmen overlap the other sets are eliminated but one, where A overlaps E and D, showing that all healthy Englishmen are healthy as well as strong—facts that we already know from the first three premisses. I overlooked this region and mistakenly thought that the premisses proved that no Englishmen exist! It is possible, I suppose, that Carroll overlooked the same region, because he adds in his letter that the conclusions surprised and bothered him. Otherwise, I find nothing surprising or bothersome about them.

4

Books and Articles

Three books by Carroll deal entirely with recreational mathematics: *The Game of Logic, Pillow-Problems,* and *A Tangled Tale. The Game of Logic* introduced Carroll's original method of diagramming syllogisms. The first edition, privately printed in 1886, had an inserted envelope containing a card on which two diagrams were printed, and nine circular cardboard counters. Four red counters were used to mark on a diagram the regions known to contain members of a set, and five gray counters were used to mark empty regions. After properly placing the counters to represent the two premisses of a syllogism, one can then determine what conclusions could be drawn by inspecting the pattern. (Two "prim-misses," we are told in *Sylvie and Bruno*

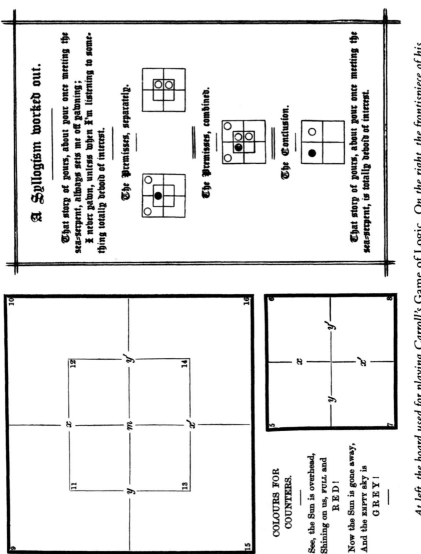

At left, the board used for playing Carroll's Game of Logic. On the right, the frontispiece of his Symbolic Logic, showing how counters are used on the board for solving a syllogism.

(Chapter 18), of a "sillygism" generate a "delusion.") A revised edition of *The Game of Logic,* also including the card and counters, was published by Macmillan in 1887.

Carroll's technique for diagramming logic statements was based on what mathematicians call Venn circles, introduced by John Venn in his *Symbolic Logic* (1881). Carroll elaborated on his square version of Venn circles in a more technical work also titled *Symbolic Logic* (1896). It was meant to be the first of three volumes on logic. Carroll never completed the last two, but galleys for the second volume were discovered by William W. Bartley, III, who reprinted them in his book *Lewis Carroll's Symbolic Logic* (1977). A definitive study of Carroll's work on logic, the book has done much to raise Carroll's reputation as a mathematician who made significant contributions to formal logic. A Dover paperback combines *Symbolic Logic* with *The Game of Logic.* At least one modern textbook on logic, *Reason and Argument* (1976), by P.T. Geach, uses Carroll's square diagrams rather than the customary Venn circles.

Carroll intended *The Game of Logic* for children, though I know of no record of any child finding the game intriguing. On the contrary, Irene Barnes, in her autobiography *To Tell My Story* (1948), recalls her frustration as a young girl when Carroll tried to teach her how to solve syllogisms with his diagram and counters. "Dare I say this made the evening rather long, when the band was playing outside on the parade, and the moon shining on the sea?"

Irene, then in her early teens, stayed a week with Carroll at the seaside resort of Eastbourne. In his diary (August 17, 1887), Carroll calls her a "charming guest" and records taking her back to London on August 23. *The Game of Logic* is dedicated to another child-friend, a niece of Henry Holiday, who illustrated *The Hunting of the Snark.* The dedication is an acrostic poem, the second letters of each line spelling "Climene Mary Holiday."

Pillow-Problems (1893) consists of seventy-two original puzzles, most of them not easily solved. The book's subtitle is *Thought Out During Sleepless Nights.* For the book's second edition he changed the last two words to "Wakeful Hours" so readers would not think he suffered from chronic insomnia. A new preface was added to the fourth edition (1895). The book was intended as Part II of what Carroll called *Curiosa Mathematica.* Part I, *A New Theory of Parallels,* was too serious to be called recreational, even though it is written with the usual Carrollian humor.

The most interesting puzzles in *Pillow-Problems* concern probability. The first such, Problem 5, is simple to state but extremely confusing to analyze correctly:

> A bag contains one counter, known to be either white or black. A white counter is put in, the bag shaken, and a counter drawn out, which proves to be white. What is now the chance of drawing a white counter?

As Carroll writes, one is tempted to answer ½. Before the white counter is withdrawn, the bag is assumed to hold, with equal probability, either a black or white counter, or two white counters. If the counters in the bag are black and white, a black counter will remain after the white one is taken. If the counters are both white, a white counter will remain after a white one is drawn. Because the two states of the bag are equally probable, it seems that after a white counter is taken, the remaining counter will be black or white with equal probability.

Carroll claims correctly that this argument, while intuitively plausible, is dead wrong, and that the answer is 2/3. To understand why he is correct, it is necessary to analyze all possible outcomes of the drawing, including the one Carroll has cunningly hidden in plain sight. The two counters can be labeled 1 and 2, but since 2 may be either black or white, we can call these alternatives 2_b and 2_w. (Since 1 is known to be white, it can be simply 1.) There are thus four possible outcomes of our first drawing:

1. Counter 1 is drawn, leaving 2_w in the bag;
2. Counter 1 is drawn, leaving 2_b in the bag;
3. Counter 2_w is drawn, leaving 1 in the bag;
4. Counter 2_b is drawn, leaving 1 in the bag.

Each of these outcomes is equally likely. Note that there is a ¼ chance of drawing a black counter (outcome 4), and also a ¼ chance that a black counter remains in the bag (outcome 2). However, Carroll has told us that outcome 4 didn't occur—the counter that is drawn out "proves to be white." This means we are left with the three outcomes in which a white counter is drawn, and of these, two involve a white counter remaining in the bag. The somewhat surprising answer, therefore, is 2/3.

The problem is easily modeled with playing cards. Shuffle a deck, spread it face down, and remove a card without looking at its face. Beside it place face down a card you know to be red. Turn your back while a friend mixes the positions of the two cards. Turn around and put a finger on one card. The chance it is red is ¾, and the chance the other card is red is also ¾. Turn over the card you are touching. If it is black, the other card *must* be red. If it is red, the probability the other card is red goes down to 2/3.

The book's last problem, number 72, has been the subject of much controversy.

A bag contains 2 counters, as to which nothing is known except that each is either black or white. Ascertain their colours without taking them out of the bag.

Here is Carroll's surprising answer:

We know that, if a bag contained 3 counters, 2 being black and one white, the chance of drawing a black one would be $^2/_3$; and that any *other* state of things would *not* give this chance. Now the chances, that the given bag contains (α) BB, (β) BW, (γ) WW, are respectively, ¼, ½, ¼.
Add a black counter.
Then the chances, that it contains (α) BBB, (β) BWB, (γ) WWB, are, as before, ¼, ½, ¼.
Hence the chance, of now drawing a black one,
$$= ¼·1 + ½·^2/_3 + ¼·^1/_3 = ^2/_3.$$
Hence the bag now contains BBW (since any *other* state of things would *not* give this chance).
Hence, before the black counter was added, it contained BW, i.e. one black counter and one white.

The proof is so obviously false that it is hard to comprehend how several top mathematicians could have taken it seriously and cited it as an example of how little Carroll understood probability theory! There is, however, not the slightest doubt that Carroll intended it as a joke. He answered all thirteen of the other probability questions in his book correctly. In the book's Introduction he gives the hoax away:

If any of my readers should feel inclined to reproach me with having worked too uniformly in the region of Common-place, and with never having ventured to wander out of the beaten tracks, I can proudly point to my one Problem in 'Transcendental Probabilities'—a subject in which, I believe, *very* little has yet been done by even the most enterprising of mathematical explorers. To the casual reader it may seem abnormal, and even paradoxical; but I would have such a reader ask himself, candidly, the question "Is not Life itself a Paradox?"

It was characteristic of Carroll that he ended his book with a choice specimen of Carrollian nonsense.

Carroll's diary entry for March 1, 1875, says that he planned to call *Curiosa Mathematica's* third volume *Alice's Puzzle Book.* He mentions that he asked Tenniel to draw a frontispiece, and Tenniel agreed. Apparently the frontispiece was never drawn. Later, when Carroll changed the book's title to *Original Games and Puzzles,* he hoped to have it illustrated with "fairy pictures" by Gertrude Thomson. Alas, the manuscript was never written.

Carroll's nephew, Stuart Dodgson Collingwood, in *The Lewis Carroll Picture Book* (reprinted by Dover under the title *Diversions and Digressions*), includes fragments of puzzle manuscripts that Carroll may have intended for his book of original games and puzzles. They include methods of multiplying and dividing large numbers, and two geometrical fallacies, not original with Carroll, which he probably did not plan to publish. One is a flawed proof that all triangles are isosceles, the other an equally deceptive proof that a right angle equals an obtuse angle. Both of these classical fallacies arise when a point where two straight lines intersect is misplaced.

A Tangled Tale (1885) consists of ten mathematical puzzles. Carroll calls them "knots," taking the noun from Alice's remark to the Mouse in the third chapter of *Alice in Wonderland:* "A knot! Oh, do let me help to undo it!" The knots first ran in *The Monthly Packet* between April 1880 and November 1884. The book is included in the Modern Library edition of Carroll's writings, and Dover has a paperback combining it with *Pillow-Problems.* Illustrations were provided by

Arthur Frost. In his diary entry for July 10, 1885, Carroll says that Frost refused to redraw his art (as usual, Carroll found many things not to his liking), but he decided to accept six of the pictures.

Seven puzzles that Carroll wrote for child-friends, each presented in verse, appeared in *Aunt Judy's Magazine* (December 1870) under the title "Puzzles From Wonderland." The answers, also in verse, ran the following month under the by-line of "Eadgyth." Edith's identity remains unknown. The verse is too inferior to have been written by Carroll.

Most of the seven puzzles rely on word play. For example:

John gave his brother James a box:
About it there were many locks.

James woke and said it gave him pain;
So gave it back to John again.

The box was not with lid supplied,
Yet caused two lids to open wide;

And all these locks had never a key—
What kind of a box, then, could it be?

The answer is that John gave James a box on the head. In another riddle, the Sun says to the Moon, "You're a Full Moon." Why was the Moon angry? Because she took the word "full" to be "fool."

Here is another riddle:

Dreaming of apples on a wall,
    And dreaming often, dear,
I dreamed that, if I counted all,
    —How many would appear?

The answer is ten, obtained by partitioning "often" to spell "of ten."

Not all the puzzles depend on puns. A man saws a two-pound stick into eight pieces of equal weight. How much does each piece weigh? The answer is a trifle less than a fourth of a pound because some weight is lost as sawdust. The seven puzzles are given in the Modern Library edition of Carroll's writings, though for some reason the verse couplet that answers the riddle about the apples is replaced by the meaningless word "Ten." The book's dedicatory poem, omitted from the Modern Library edition, is an acrostic, the second letters of each line spelling "Edith Rix."

Carroll published two amusing papers in the philosophical journal *Mind,* each about a paradox that can, in a wide sense of the term, be called recreational. Even if they are taken seriously, Carroll's inimitable way of presenting them is certainly entertaining.

"A Logical Paradox" (*Mind,* July 1894) discusses in fictional form a seeming contradiction in what logicians call the propositional calculus. Carroll first mentioned the paradox in his diary (March 31, 1894):

> Have just got printed, as a leaflet, *A Disputed Point in Logic—*
> the point Prof. Wilson and I have been arguing so long.
> This paper is *wholly* in his own words, and puts the point
> very clearly. I think of submitting it to all my logical friends.

"Prof. Wilson" was John Cook Wilson (1849–1915), a classical scholar who taught logic at Oxford (in those days logic was considered part of philosophy, not a subject to be taught by mathematicians as it is today). Carroll and Wilson argued for years over the "disputed point." The disagreement centered on how to interpret what today is called the binary relation of implication, but in Carroll's day the relation was called a "hypothetical." These are statements of the form "If A is true, then B is true." Carroll and Wilson exchanged numerous letters about hypotheticals, some of them mentioned in Carroll's diary. In light of today's clear understanding of implication, as used in the propositional calculus, their dispute is as trivial as it is funny, but in Carroll's time the meaning of implication was none too clear. It was not until seven years after Carroll's death that Wilson capitulated by writing a note in *Mind* (Volume 14, April 1905, pages 292–93) admitting that Carroll had been right all along!

Wilson's recantation, signed only with the initial W, correctly concludes: "The fallacy then is a mere verbal one, caused by a misunderstanding of what is exactly meant by saying that the proposition: 'If Allen is out, Brown is in' is a consequence of the proposition 'Carr is out'. . . . Mr. Dodgson's argument makes no wrong use of it and is, so far, quite sound. . . . It is curious what slips can be made in formal reasoning. No one seems safe from them."

Carroll's four-page anonymous pamphlet, "A Disputed Point in Logic" (March 1894), was his first printed version of the paradox. It took the form of a dialogue between Nemo (who represents Wilson) and Outis (Carroll). Both names are Greek for "nobody." (In Homer's *Odyssey*, remember, Ulysses calls himself Outis to confuse the one-eyed giant Polyphemus, whom he has just blinded.)

In April 1894, Carroll rewrote the pamphlet and published it under the same title as before. For A, B, and C he substituted three men: Allen (A), Brown (B), and Carr (C). Unlike the previous pamphlet, the revision contains no quotes from Wilson's letters. The dialogue is entirely in Carroll's own words.

The three men live in the same house. A man is either in the house or out. "Out" is equivalent to "true," "in" is equivalent to "false." Stated in the propositional calculus, there are two axioms:

1. If A is true, B is true.
2. If C is true, and if A is true, B is not true.

Because the conclusion of 2 contradicts the conclusion of 1, it seems that C cannot be true.

We now put the paradox in terms of the men being in or out of their house.

1. If Allen is out, Brown is out.
2. If Carr is out, and if Allen is out, Brown is in.

Must we conclude, from the contradiction, that Carr cannot be out?

Several other versions of the problem, written by Carroll but unpublished, are given in Bartley's *Lewis Carroll's Symbolic Logic* (page 449ff).

On May 4, 1894, Carroll wrote in his diary: "Yesterday I wrote out 'Allen and Co.' paradox in the form of dialogue for *Mind.*" After this appeared in the July 1894 issue, Carroll reprinted it as a four-page document to distribute to friends. Eventually he planned to include it in his projected, but never published, second volume of *Symbolic Logic*. Galley proofs of this book survive, owned by the Christ Church Library, and are reproduced in Bartley's volume.

The roles of Nemo and Outis are now taken by Uncle Joe (Wilson) and Uncle Jim (Carroll). Their nephew Cub, a fifteen-year-old boy, narrates what is essentially a short-short story. It can be found in Bartley's book, in Collingwood's *The*

Lewis Carroll Picture Book, in John Fisher's *The Magic of Lewis Carroll,* and in R.M. Eaton's *General Logic.* It is not included in the Modern Library's mistitled *The Complete Works of Lewis Carroll.*

Allen, Brown, and Carr have now become three barbers who occupy the same shop. The two axioms are:

1. If Allen goes out, Brown always goes with him.
2. All three are never out of the shop at the same time.

Obviously there is nothing contradictory about the two axioms. Carr can go out whenever he pleases. However, suppose we assume that Carr goes out, then Allen goes too. Axiom 1 tells us that if Allen goes out, Brown must also go out. But this violates the second axiom by leaving the shop unattended! Uncle Joe (Wilson) maintains that we have here a *reductio ad absurdum,* which proves that Carr can never go out. On the other hand, as Uncle Jim (Carroll) insists, Carr obviously can go out without violating either axiom. Hence the paradox.

Uncle Jim is clearly right. The fallacy lies in assuming that if Carr goes out, it is possible for Allen to go with him. This leads to a contradiction, and nonsense results. Put another way, if Carr goes out, axiom 2 prevents Allen from going with him. Brown can go out whenever he pleases, either alone or with Allen or Carr, but not with both. If Carr goes out, Allen must stay in. If Allen is out, and if Brown is out, Carr must stay in.

Many correct resolutions of the paradox have been published. John Venn, in *Symbolic Logic* (pages 442–43), analyzes it in Boolean notation. Alfred Sidgwick and W.E. Johnson each discussed it in *Mind* (Volume 3, October 1894, pages 582–83) and the following year in *Mind* (Volume 4, January 1895, pages 143–44). E.E.C. Jones analyzed the paradox in *Mind* (Volume 14, January 1905, pages 146–48). Bertrand Russell covers it briefly in *The Principles of Mathematics* (1903), and there are resolutions by Irving Copi, Arthur Burks, and others. R.B. Braithwaite considered the paradox in "Lewis Carroll as Logician," in *The Mathematical Gazette* (July 1932), and Warren Weaver discusses it in "Lewis Carroll Mathematician," in *Scientific American* (April 1956).

Letters from Weaver and Alexander Morris also appeared in the June 1956 issue. "Is it possible," Morris asks, "that this is really a serious problem for the professional logician? . . . [It is] a perfectly simple set of facts. . .which any reasonably bright child of ten should be able to manipulate."

Carroll's story ends with Cub and his uncles entering the barbershop, where "we found—"The dash ends the tale, leaving open whether they found Carr inside, as Uncle Joe believed he had to be, or outside, as Uncle Jim was sure he could be.

Carroll's understanding of the propositional calculus, as Bartley makes clear in his book, was suprisingly modern. Bartley quotes the following passage from the December 21, 1894, entry in the diary—a passage omitted in Green's edition:

> My night's thinking over the very puzzling subject of 'Hypotheticals' seems to have evolved a new idea—that there are *two* kinds, (1) where the Protasis is *in*dependent of the Hypothetical, (2) where it is dependent on it.

The passage reveals that Carroll understood that implication can be interpreted in two radically different ways. He never followed up on this notion, but he clearly grasped the distinction between what logicians call "material implication," as used in the propositional calculus, and "strict implication," as in various modal logics, which assume that in the statement "A implies B," B is causally related to A.

"What the Tortoise Said to Achilles" (*Mind,* December 1894)—Carroll later reprinted it as a four-page pamphlet— deals with a far from trivial question. It amounts to this. In logic and mathematics you cannot prove a theorem except within a formal system based on a set of posits or assumptions. But are the assumptions true? To prove them you have to make additional assumptions, and to prove *those* assumptions requires still further posits. You thus seem to be trapped in an infinite regress. Deductions can never reach absolute certainty. You are forced to stop at some point and accept a set of posits as true by an act of faith.

The argument goes back to Agrippa, an ancient Greek skep- tic who claimed that nothing in mathematics is certain because every proof requires a proof that the proof is valid, and so into the regress. Bertrand Russell thought that this paper was Carroll's most important contribution to logic. Ways of avoid- ing the regress have been defended in dozens of papers by phi- losophers, logicians, and mathematicians. You'll find a good discussion of this in Appendix C of Bartley's book, with many references, including several papers on the topic by Bartley.

A subtle form of an endless regress arises from Kurt Gödel's revolutionary work on undecidable statements in any formal system complicated enough to include arithmetic. There nec- essarily will be statements that cannot be proved true without adding a new posit that enlarges the system. But in the en- larged system, undecidable statements are also unavoidable, and so on up the endless ladder of meta-systems. There is therefore a sense in which absolute certainty about all statements in a formal system is forever beyond the mathematician's grasp.

Carroll presents his argument in the form of an amusing dialogue between Achilles and a Tortoise. One of Zeno's paradoxes of motion seemed to show that Achilles can never catch the Tortoise because when he arrives at the spot where the Tortoise had been, the reptile has crawled ahead. And when Achilles runs *that* distance, the animal has moved ahead again. (Think of Achilles and the Tortoise as two points moving along a straight line.) Indeed, if the points move in discrete steps— first Achilles, then the Tortoise—with equal pauses after each move, the distances between them get smaller and smaller but never totally vanish, and Achilles cannot catch the Tortoise in a finite period of time. Only when the two motions are continuous does Achilles overtake the beast.

In Carroll's dialogue, Achilles has just won the race and is sitting on the Tortoise's back. Achilles tries his best to defend the certainty of deduction, only to have the Tortoise raise an endless series of demands that his proofs be proved. The paper ends with two dreadful puns:

> The Tortoise was saying, "Have you got that last step written down? Unless I've lost count, that makes a thousand and one. There are several millions more to come. And *would* you mind, as a personal favour—considering what a lot of instruction this colloquy of ours will provide for the Logicians of the Nineteenth Century—*would* you mind adopting a pun that my cousin the Mock-Turtle will then make, and allowing yourself to be re-named *Taught-Us?*"
>
> "As you please!" replied the weary warrior, in the hollow tones of despair, as he buried his face in his hands. "Provided that *you,* for *your* parts, will adopt a pun the Mock-Turtle never made, and allow yourself to be re-named *A Kill-Ease!*"

"If 6 cats kill 6 rats in 6 minutes, how many will be needed to kill 100 rats in 50 minutes?" Carroll analyzed this popular conundrum in a short article, "Problem: Cats and Rats," in

The Monthly Packet (February 1880). He shows that the problem, like so many others of a similar sort, is too ambiguous to permit a solution. One must know the exact procedure by which the rats are killed. Carroll reduces the question to absurdity by asking, "If a cat can kill a rat in a minute, how long would it be killing 60,000 rats? Ah, how long, indeed! My private opinion is that the rats would kill the cat."

Carroll had an intense dislike of arithmetical problems given in a story form that rendered them impossible to solve. Bishop T.B. Strong, in "Mr. Dodgson: Lewis Carroll at Oxford," an essay reprinted in Morton Cohen's anthology, *Lewis Carroll: Interviews and Recollections* (1989), recalls Carroll asking a class this question: "If it takes 10 men so many days to build a wall, how long would it take 300,000?" Any answer giving a short period of time, Carroll would point out, is absurd. "The wall would go up like a flash of lightning, and most of the men could not have got within a mile of it."

5

Miscellaneous Amusements

A mong Carroll's papers were a number of manuscripts dealing with puzzles and other mathematical curiosities that interested him, and which he may have planned to put into books. You'll find these items in Stuart Collingwood's *The Lewis Carroll Picture Book,* in John Fisher's *The Magic of Lewis Carroll,* in Edward Wakefield's *Lewis Carroll's Games and Puzzles,* and in other books and articles.

The items are, for the most part, not original with Carroll. They include a river-crossing problem about four men and their wives, a magic square to be produced by arranging British postage stamps in a three-by-three matrix, and a combinatorial problem similar to a river-crossing puzzle but involving a pulley with a weight on one end and a basket on the other. The task is to get a queen, her daughter, and her son down from a tower where they have been held captive.

One item is a curiosity involving British money:

> Put down any number of pounds not more than twelve, any number of shillings under twenty, and any number of pence under twelve. Under the pounds put the number of pence, under the shillings the number of shillings, and under the pence the number of pounds, thus reversing the line.
>
> Subtract.
> Reverse the line again.
> Add.
>
> Answer, £12 18s. 11d., *whatever* numbers may have been selected.

Another item deals with what mathematicians call a "cyclic number":

> A MAGIC NUMBER.
> 142857.
> 285714 twice that number.
> 428571 thrice that number.
> 571428 four times that number.
> 714285 five times that number.
> 857142 six times that number.
>
> Begin at the "1" in each line and it will be the same order of figures as the magic number up to six times that number, while seven times the magic number results in a row of 9's.

Cyclic numbers are the repeated sequences of decimals obtained by dividing 1 by certain primes, in this case, by 7. There is a sizable literature on their amazing properties, and magicians have invented many clever tricks based on 142857. The number is mentioned in Carroll's diary (October 6, 1897) and is probably the basis of the "number repeating puzzle" in an entry for January 26 of the same year. The cyclic property of 142857 was published anonymously in *Chatterbox* (February 1897) where Carroll may have first encountered it, or perhaps contributed it himself, although the curiosity had long been known to number theorists.

In *Lewis Carroll: Interviews and Recollections,* Morton Cohen reprints an article by Edward Gordon Craig, son of the actress Ellen Terry. He recalls as a boy hearing Carroll show a river-crossing problem to his mother:

> We were still living at 33 Longridge Road, Earl's Court. Here . . . I saw Lewis Carroll once. He had called to see E[llen] T[erry] at about six o'clock. She was asleep—but about to get up, so as to go to act at the theatre. I can see him now, on one side of the heavy mahogany table—dressed in black, with a face which made no impression on me at all. I on the other side of the heavy mahogany table, and he describing in detail an event in which I had not the slightest interest—'How five sheep were taken across a river in one boat, two each time—first two, second two—that leaves one yet two must go over'—ah—he did this with matches and a matchbox—I was not amused—so I have forgotten how these sheep did their trick.

In a letter to a child-friend (*The Letters of Lewis Carroll,* edited by Morton Cohen, page 300) Carroll speaks of a trick involving "two thieves and five apples." Because only magic buffs like me know this ancient trick, let me explain. You need seven identical small objects, such as pennies, matches, or buttons. Let's assume you use matches. Put five in a row on the table to represent five apples in a basket outside a farmer's barn. A match held in each fist represents a thief.

The thieves plan to steal the apples. So saying, pick up the five matches one at a time, starting with your right hand and alternating hands. The thieves see the farmer opening the door of his house so they quickly replace the apples, one at a time, alternating hands as before. This time start replacing with your *left* hand. Unknown to your viewers, the left fist is now empty.

The thieves, you continue, hide behind the barn until the farmer goes back inside his house. They then come back and take the apples again. This time start the pickup with your *right* hand. The thief in your left hand is puzzled. How is it that he has only *one* apple—open your left hand and toss two matches to the table—whereas the other thief has *four*. Open your right hand and toss five matches to the table.

Lancelot Robson's article "Give My Love to the Children" (*Reader's Digest,* February 1953) is also reprinted in Cohen's anthology. Robson recalls a children's party he attended as a youngster at which Carroll performed two magic tricks with numbers. One involved an old lightning-calculation method (described in Chapter 2) of quickly obtaining the sum of five four-digit numbers. The other stunt is still an effective one to show children who have access to a hand calculator. Put 12345679 (the 8 is omitted) in the readout and ask the child to name his favorite digit. In your head, multiply the digit by 9, and then ask the child to multiply 12345679 by the product. He will get a row of nine repetitions of his named digit. For example, suppose he said 5. Five times 9 is 45. When he multiples 12345679 by 45, the result is 555555555. The trick works because 9 times 12345679 is 111111111.

Carroll liked to carry with him a variety of mechanical puzzles, and many of his letters speak of giving them to children to solve. He also delighted in self-working magic tricks, like the thieves and apples, that depend on mathematical principles rather than sleight-of-hand. Cohen's anthology contains a recollection by Dorothy Birch of Carroll showing how a ha'penny could be pushed through a much smaller hole in a sheet of paper. This is done by folding the sheet across the hole and allowing the paper to bend without tearing as the coin is pushed through. The trick works provided the diameter of the coin is smaller than half the circumference of the circular hole.

In the biography of his uncle, Stuart Dodgson Collingwood quotes from a manuscript found among Carroll's effects:

> A is to draw a fictitious map divided into counties.
> B is to colour it (or rather mark the counties with *names* of colours) using as few colours as possible.
> Two adjacent counties must have *different* colours.
> A's object is to force B to use as *many* colours as possible. How many can he force B to use?

The answer is four. The first player can force B to use four colors with this ridiculously simple map:

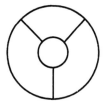

On more complicated maps it is not obvious whether a fifth color is required. Actually, every map on the plane or on a sphere can always be colored with four colors so that no two regions of the same color share a boundary. This was only a conjecture in Carroll's day, so he could not have known with certainty whether the answer to his question was four or five. Even today a tiny uncertainty lingers over the validity of the proof of what is known as the "four-color map theorem." The proof occupies such a horrendous mass of computer printouts that there could be a subtle flaw that no one has yet detected. Even if the theorem is true, as it almost certainly is, it still awaits a simple proof that does not require hours of computer time.

Viscount John Allsebrook Simon, in Derek Hudson's *Lewis Carroll* (1954), recalls some puzzles Carroll liked to spring on friends. One was the classic problem about two glasses, one of water, the other of brandy, each holding fifty spoonfuls. A spoonful of brandy is put in the water, the liquid is stirred, and then a spoonful of the mixture is transferred back to the brandy. Which glass now contains the most of the other liquid? As Simon explains:

> The answer, of course, is easy enough to work out, for the spoonful of the mixture will consist of 1/51 parts of brandy and 50/51 parts of water, so on the whole transaction 50/51sts of brandy has been transferred from the first tumbler to the second, and 50/51sts of water from the second tumbler to the first. But Lewis Carroll then observed that it was quite unnecessary to work out these fractions. You started with a tumbler containing 50 spoonfuls of brandy and at the end this tumbler still contained 50 spoonfuls, neither more nor less. Whatever it had lost in brandy it had gained in water, and as there had been no spilling the quantities were equal.

Carroll could have added, and perhaps did, that the glasses need not be the same size, that stirring is not necessary, and that any amounts of liquid can be transferred back and forth as many times as desired, provided that at the finish each glass holds the same amount as at the start.

The figure on the next page reproduces a maze Carroll drew for his youthful magazine *Mischmasch*. (Incidentally, complete runs of this magazine and another called *The Rectory Umbrella* are available in a single Dover paperback, with a foreword by Florence Milner.) In his biography of Carroll, Collingwood tells how Carroll, as a boy, once traced out in the snow "a maze of such hopeless intricacy as almost to put its famous rival at Hampton Court in the shade."

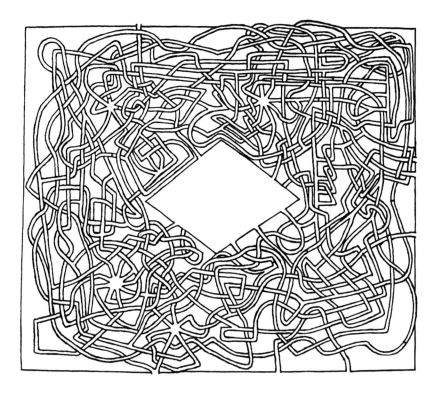

A maze drawn by Lewis Carroll in his early twenties. The object is to find your way out of the central space. Paths cross over and under one another, but are occasionally blocked by single-line barriers.

Two intriguing mathematical curiosities were published by a teenage Carroll in *The Rectory Umbrella*. One had to do with the difficulty, which was very real at the time, of deciding where to alter clocks during trips around the earth. International date lines were not proposed until 1878, and were put into effect in

1884. In Carroll's typically amusing discussion of deciding what day it was, he correctly predicted that

> some line would have to be fixed where the change should take place, so that the inhabitants of one house would wake and say, "Heigh-ho, Tuesday morning!" and the inhabitants of the next (over the line), a few miles to the west would wake a few minutes afterwards and say, "Heigh-ho! Wednesday morning!" What hopeless confusion the people who happened to live *on* the line would be in, is not for me to say. There would be a quarrel every morning as to what the name of the day should be.

Carroll later published a letter about this in *The Illustrated London News* (April 18, 1857). In 1860, he lectured on the problem to the Ashmolean Society.

The other curiosity was also about time. In an 1849 letter to a sister, later printed in *The Rectory Umbrella*, Carroll asked which clock is more accurate—one that is right once a year, or a clock that doesn't run at all? The answer is the latter because it is correct twice every day!

Suppose that the stopped clock shows 8 o'clock.

> You *might* go on to ask. 'How am I to know when eight o'clock *does* come? My clock will not tell me.' Be patient, reader: you know that when eight o'clock comes your clock is right: very good; then your rule is this, keep your eye fixed on your clock, and *the very moment it is right* it will be eight o'clock. 'But—' you say. There, that'll do, reader: the more you argue, the farther you get from the point, so it will be as well to stop.

6

Doublets

Doublet tasks consist of changing one word to another by altering single letters at each step to make a different word. The two words at the beginning and end of such a chain must, of course, be the same length, and they should be related to each other in some obvious way. They must not have identical letters in the same positions. All words in the chain should be common English words, proper names excluded. A "perfect" solution has a number of steps equal to the number of letters in each word. For example: COLD, CORD, CARD, WARD, WARM. If a perfect chain is not possible, the best solution is the shortest chain. For playing doublets as a game with two or more players, Carroll invented a set of scoring rules to determine who wins.

The first diary mention of the game is on March 12, 1878, when Carroll reports teaching "Word Links" (his original name for the game) to guests at a dinner party. He had invented the game, he tells us in a pamphlet, on Christmas day, 1877, for two little girls who "found nothing to do."

Carroll's hand-lettered *Word-Links: A Game For Two Players, or a Round Game,* written in April 1878, is reprinted here for the first time. Later that year he printed a revised version as a four-page pamphlet. Starting with the March 29, 1879, issue of *Vanity Fair,* Carroll contributed a series of articles on doublets. The first piece is reprinted here, followed by an article announcing a doublet competition, and a third article giving a new method of scoring. Also included is a leaflet giving the doublets set for six previous contests, and announcing the next contest on August 2.

In 1879, Macmillan gathered the *Vanity Fair* articles into a 39-page book, with red cloth covers, titled *Doublets: A Word Puzzle.* An 1880 second edition was enlarged to 73 pages. *The Lewis Carroll Picture Book* reprints part of this edition. Later that same year Macmillan published a third edition, revised and enlarged to 85 pages.

Carroll took the name doublets from a line of the witches' incantation in Shakespeare's *Macbeth:* "Double, double, toil and trouble"—a line Carroll placed on the title page of his book.

On May 11, 1885, Carroll mentions in his diary that he has extended his list of seven-letter word pairs that can be linked together to more than 500.

Doublets became a parlor craze in London and have been a much practiced form of word play ever since. They have been called by other names, such as "word ladders," and (in Vladimir Nabokov's novel *Pale Fire*) "word golf." Enormous energies have been expended on finding the shortest ladders for a given pair of words. Computer software containing all English words is now obtainable, and programs have been written for finding minimum chains in a few seconds. The task is equivalent to finding the shortest routes connecting two points on a graph—a task closely related to what are called Gray codes.

Donald Knuth, Stanford University's noted computer scientist, has constructed a graph on which 5,757 of the most common five-letter English words (proper names excluded) are represented by points, each joined by a line to every word to which it can be changed by altering just one letter. The graph has 14,135 lines. Once it is in a computer's memory, programs can be written that will determine in a split second the shortest word ladder joining any two words on the graph. Knuth found three-letter words too simple and six-letter words less interesting because not too many can be connected.

Most pairs of five-letter words on Knuth's list can be joined by ladders. Some—Knuth calls them "aloof" words because one of them is the word *aloof*—have no neighbors. The graph has 671 aloof words, including *earth, ocean, below, sugar, laugh, first, third,* and *ninth*. Two words, *bares* and *cores,* are connected to 25 other words; none are connected to a higher number. There are 103 word pairs with no neighbors except each other, such as ODIUM–OPIUM and MONAD–GONAD. Knuth's 1992 Christmas card featured the smallest ladder (eleven steps) that changes SWORD to PEACE using only words found in the Bible's Revised Standard Edition.

I have taken the above information from the eight pages devoted to doublets in the first chapter of Knuth's book *The Stanford GraphBase* (Addison Wesley, 1993). Knuth will cover the topic more fully in his forthcoming three-volume work on combinatorics in his classic *Art of Computer Programming* series. For hints on how to solve doublets without a computer, see his article "WORD, WARD, WARE, DARE, GAME," in *Lewis Carroll's Games* (July–August, 1978).

It has been pointed out that doublets resemble the way in which evolution creates a new species by making small random changes in the "genes" that are intervals along the helical DNA molecule. Carroll himself, although a skeptic of Darwin's theory, evolved MAN from APE in six steps:

APE
ARE
ERE
ERR
EAR
MAR
MAN.

When I gave this solution in a *Scientific American* column on mathematical games (the column is reprinted as Chapter 4 in my *New Mathematical Diversions*), two readers produced a shorter solution:

APE
APT
OPT
OAT
MAT
MAN.

In a letter on March 12, 1892 (see Morton Cohen's *The Letters of Lewis Carroll*, Volume 2, page 896), Carroll added a rule that allows one to rearrange the letters of any word, counting this as a step. With such increased freedom, he pointed out, many impossible doublets, such as changing IRON to LEAD, can be achieved: IRON, ICON, COIN, CORN, CORD, LORD, LOAD, LEAD.

It is difficult but not impossible for a word chain to form a sentence. In *Vanity Fair* (July 26, 1879), one of Carroll's doublets asked "WHY is it better NOT to marry?" To change WHY to NOT he added this proviso:"The chain made [WHY to NOT] . . . should embody the following observation: that lovers, during the temporary insanity of courtship, too often fail to recognize the grave prudential reasons which should deter them from taking this fatal step." Here is Carroll's clever solution: WHY, WHO, WOO, WOT, NOT.

The mathematician and science fiction writer Rudy Rucker has likened doublets to a formal system. The first word is the given "axiom." The steps obey "transformation rules," and the final word is the "theorem." One seeks to "prove" the theorem by the shortest set of transformations.

Many papers on doublets have appeared in the journal *Word Ways*, a quarterly devoted to linguistic amusements. An article in the February 1979 issue explored chains that reverse a word, such as TRAM to MART, FLOG to GOLF, LOOPS to SPOOL, and so on. The author asks if an example can be found using six-letter words. Is there a closed chain, I wonder, that changes SPRING to SUMMER to AUTUMN to WINTER, then back to SPRING? If so, what is the shortest solution? The task is probably impossible because "AUTUMN" seems to be an "aloof" word, but, as Knuth points out, it may be solvable if one is allowed to rearrange letters at each step.

A.K. Dewdney, in a "Computer Recreations" column in *Scientific American* (August 1987), calls a graph connecting all words of *n* letters a "word web." He shows that all two-letter words are easily joined by such a web, and asks if anyone can construct a complete word web for three-letter words.

Mathematician Ian Stewart, in *Nature's Numbers* (1995, pages 41–43), proves an interesting theorem about doublets. If the first word has a vowel in a certain position, and the final word has a vowel in a different position, there must be an intermediate word with at least two vowels. His example is changing SHIP to DOCK.

WORD-LINKS.

A Game for two Players, or a round Game.

The principal feature of this game consists in the linking together of words, so that any two conse-cutive words may differ in one letter only. Such a series of words is called a "Chain". The simp-lest form of Chain is where Two words are given, and the Chain is so made as to have the given words at the two ends (such as "Head, Heal, Teal, Tell, Tall, Tail".) The two given words are called a "Doublet", and the words intro-duced to make the Chain are called "Links". A more elaborate form of Chain is where there are three or more words given, (such as "Ale, Ice, Arm, Ant") and the Chain is so made as to have two of them at the ends, and the others in the Chain, (such as "Ale, Are, Ire, Ice, Ace, Act, Ant, Art, Arm"). The given words are called "Jewels", and all that have chain on both sides of them are said to be "set". If Links can be found to unite the two ends (such as "Aim, Ail, All," which would unite "Arm" to "Ale") the Chain is called a "Neck-lace", and all the Jewels in it are then said to be "set". The above Necklace may be written in this form :—

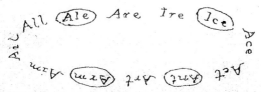

In making a Chain or Necklace, it is not allowable to use a word twice.

We will now suppose that each of the Players is provided with a large sheet of paper to work on, a number of slips

to write out Chains on, a limited stock of pa-
-tience, and an unlimited stock of good-tem-
-per. We proceed to the

Rules for 2 Players.

One player opens a book at random and
hands it to the other, who selects, at the open
place, 3 words of 3 letters each, 3 of 4, and
3 of 5 : he then opens it in another place
and returns it to the first Player, who se-
-lects 9 words in the same manner. These
10 words are written down by each Play-
-er, and are the "Jewels."

The first Player then opens the book
again and hands it to the other, who se-
-lects a word of 3, or 4, or 5 letters, as
may be agreed : he then opens it in an-
-other place and returns it to the first
Player, who selects another word of the
same length. These words form the first
"Doublet."

Each Player then works this Doublet
into a Chain, which he writes out on a
slip of paper and lays face-downwards
on the table.

When both are written, the papers are
turned up, and the one who finished the
first gets 2 marks for "speed": and, if
the Chains are of different lengths, the
writer of the shortest gets so many
marks, as represent the difference, for
"brevity." Another Doublet is then selected,
and the game proceeds as before.

If one Player abandons the attempt
to make a Chain, but the other succeeds;
the successful one gets 2 marks for
"speed" and 4 for "brevity."

If both abandon the attempt, another
Doublet is selected.

A Chain which contains an inadmissi-
-ble word, or which violates any of the rules,

is called "null"

If both Players fail in making a Chain, he who first declared his to be "abandoned" gets 3 marks for "decision"; but, if both Chains be "null," neither Player gets marks.

The Player who first finishes (or abandons) his Chain may occupy his time, till the other has finished, in making "extra-Chains" of the Jewels. These are marked at the end of the game, by the following rule:— a Jewel at the end of a Chain gets as many marks as it has letters; a "set" Jewel twice as many; and every Link loses a mark. Thus, a Chain of 6 Jewels of 4 letters each, containing 30 Links, gets 10 marks: if, by adding 6 more Links, it is made into a Necklace, it gets 12.

In making a set of "extra-Chains", it is not allowable to use any Jewel twice; but a word may be used more than once as a Link, so long as it only occurs once in each Chain.

If a set of extra-Chains violate any rule, there may still be portions which can get marks, and the writer is allowed to erase the faulty portions.

We now proceed to the

Rules for a Round Game,

which are the same as for 2 Players, with the following exceptions :—

A President is chosen to score the marks: he may play himself.

The President opens a book at random, and hands it to one of the players, who selects, at the open place, a word of 3 letters, a word of 4, and a word of 5. He then does the same with another Player, and so on, till the 18 Jewels have been selected.

The first Doublet is selected in the same way As soon as any Player has finished his

Chain, he writes it on a slip of paper and hands it, face-downwards, to the President, who places the slips in a heap.

When all have handed in (or abandoned) their Chains, the President turns over the heap, and marks for "speed" by the following rule. — the first correct Chain finished gets as many marks as there are Players; the next gets one mark less, and so on, the last getting none. He then marks for "brevity" as follows: — the longest Chain gets no marks, and every shorter Chain gets as many marks as represent the difference. If any Chain be abandoned or null, it is taken as "longest" and considered to be 4 Links longer than the longest successful one.

If all fail in making a Chain, he who first declared his to be abandoned gets 3 marks for "decision", and the one who declared next gets 2.

Before beginning a game, it should be settled how many Doublets it is to consist of: six will be found a convenient number.

The following form of scoring-paper is recommended:— [N.B. The first figure in a "Doublet" column is for "speed"; the second for "brevity". "A" stands for "abandoned the first"; "a" for "abandoned second"; "O" represents "no marks".]

Date	Names	1	2	3	4	5	6	Jewels	Total
Ap.1/78	Smith	2,0	a	4,3	3,6	2,4	O	8	34
	Brown	4,4	A	3,0	a	4,8	4,4	11	45
	Jones	O	O	2,0	4,8	O	A	4	18
	Robinson	3,3	O	0,3	A	3,6	a	8	20

Ap 11. 1878   Lewis Carroll

WORD-LINKS.

A Game for two Players, or a Round Game.

A SERIES of words of the same length, where any two consecutive ones differ in one letter only (e. g. 'Head, Heal, Teal, Tell, Tall, Tail'), is called a 'Chain.' The game consists in forming Chains so as to contain two or more given words, one of which must be at each end. Two given words are called a 'Doublet'; three or more, 'Jewels'; connecting words are called 'Links.' (Thus the above Chain might have been made for the 'Doublet' 'Head, Tail,' or for the four Jewels 'Head, Teal, Tell, Tail.') A Jewel that has chain on both sides (e. g. 'Teal' in the above Chain) is said to be 'set.' If Links be found to unite the two ends (e. g. 'Hail, Hair, Heir, Hear,' which unite 'Tail' to 'Head'), the Chain is called a 'Necklace,' and *all* the Jewels in it are said to be 'set.' A Necklace must contain at least three Jewels. The above Necklace might be written thus :—

Head	Heal	*Teal*	*Tell*
Hear			Tall
Heir	Hair	Hail	*Tail*

In making a Chain or Necklace, it is not allowed to use a word twice; and no word is admissible that is not in ordinary use in good society.

2

THE RULES FOR TWO PLAYERS.

Each Player should be provided with writing-materials and an English book; and one of them should also have a scoring-paper, ruled as follows :—

NAMES.	DECLARATIONS, &c.								MARKS.	
	1	2	3	4	1	2	3	4	Jewels.	Total.
Brown.										
Jones.										

Each opens a book, and selects two words of three letters, two of four, and two of five. These twelve words are written down by each, and are the 'Jewels.'

Each again opens a book, and selects a word of three, four, five, or any other number of letters, as may be agreed. When both words are fixed on they are read out, and are the first 'Doublet.'

Each now tries to make a Chain of this. As soon as a Player has decided with how many Links he will undertake to make it, or that he will abandon the attempt, he says, '— Links,' or 'Abandoned,' as the case may be. He who does this first is marked 'I' in the first 'Declaration'-column, followed by the number he named or by the letter '*a*': the other is marked 'II' in the same way.

A Chain is reckoned as having the 'declared' number of Links, even if it really have fewer : if it have more, it is 'null.'

3

When both have 'declared,' each must at once write out his Chain (if he has not already done so); if he cannot do this, it is 'null.'

The Chains are now examined, and correct Chains are marked as follows :—He who first 'declared' the length of his Chain gets 2 for 'decision;' also the writer of the shorter gets, for 'brevity,' 3 for every Link saved. (N.B. A 'null' Chain is reckoned as being two Links longer than a correct one.) If neither be correct, he who first 'abandoned' gets 2 for 'decision.' All these numbers are entered in the first 'Marks' column.

A second Doublet is then selected, and the game proceeds as before, four Doublets making one Game.

Each Player may employ his spare time in making extra Chains or Necklaces of the Jewels. These are marked at the end of the game, thus :—A Jewel at the end of a Chain gets as many marks as it has letters; a 'set' Jewel twice as many; and every Link loses a mark. (Thus a Chain with three Jewels, of four letters each, and seven links, gets 9; if, by adding four links, it be made into a Necklace, it gets 13.)

In making extra Chains, it is not allowed to use any Jewel twice; but a word may be used more than once as a Link, provided it does not occur twice in one Chain.

If a set of extra Chains transgress any rule, there may still be portions which can get marks; and the writer is allowed to withdraw the faulty portions.

4

THE RULES FOR A ROUND GAME

are the same as the above, with the following alterations :—

Each Player should have, besides a large sheet of paper for working on, a number of slips for writing out Chains when finished.

One of the Players is chosen as President: he keeps the score, and settles all disputed points.

Each Player in turn opens a book and selects a word of three letters, one of four, and one of five; till the twelve jewels have been selected.

Two Players, named by the President, select the first Doublet.

Correct Chains are marked as follows :—He who first 'declared' the length of his Chain gets, for 'decision,' as many marks as there are Players, the next gets one mark less, and so on down to '2,' which is the lowest mark given ; also the writer of any Chain that is not the longest gets, for 'brevity,' 3 for every link saved. (N.B. A 'null' chain is reckoned as being two links longer than the longest correct one.) If none be correct, he who first 'abandoned' gets 4 for 'decision,' and the next gets 2.

The following Doublets may be useful for practice in making Chains :—

'Hare, Soup, has been done with six links; 'Tree, Wood,' with eight ; 'Pen, Ink,' with eight ; 'Castle, Butler,' with ten ; 'Mine, Coal,' with six ; 'Grub, Moth,' with twelve ; 'Quilt, Sheet,' with eighteen ; 'Bread, Toast,' with twenty-one.

March. 29. 18

A NEW PUZZLE.

THE readers of *Vanity Fair* have during the last ten years shown so much interest in the Acrostics and Hard Cases which were first made the object of sustained competition for prizes in this journal, that it has been sought to invent for them an entirely new kind of puzzle, such as would interest them equally with those that have already been so successful. The subjoined letter from Mr. Lewis Carroll—a name already immortalised as that of the author of "Alice in Wonderland" —will explain itself, and will introduce a puzzle so entirely novel and withal as interesting, that the transmutation of the original into the final word of the Doublets may be expected to become an occupation to the full as amusing as the guessing of the Double Acrostics has already proved.

In order to enable readers to become acquainted with the new puzzle, preliminary Doublets will be given during the next three weeks—that is to say, in the present number of *Vanity Fair* and in those of the 5th and 12th April. A competition will then be opened—beginning with the Doublets published on the 19th April, and including all those published subsequently up to and including the number of the 26th July—for three prizes, consisting respectively of a Proof Album for the first and of Ordinary Albums for the second and third prizes.

The rule of scoring will be as follows :—A number of marks will be apportioned to each Doublet equal to the number of letters in the two words given. For example, in the instance given below of "Head" and "Tail," the number of possible marks to be gained would be eight; and this maximum will be gained by each one of those who make the chain with the least possible number of changes. If it be assumed that in this instance the chain cannot be completed with less than the four links given, then those who complete it with four links only will receive eight marks, while a mark will be deducted for every extra link used beyond four. Any competitor, therefore, using five links would score seven marks, any competitor using eight links would score four, and any using twelve links or more would score nothing. The marks gained by each competitor will be published each week.

In order to afford space for the Doublets the publication of Hard Cases will be discontinued from and after the 17th April.

"DOUBLETS"—A VERBAL PUZZLE.

DEAR VANITY,—Just a year ago last Christmas, two young ladies—smarting under that sorest scourge of feminine humanity, the having "nothing to do"—besought me to send them "some riddles." But riddles I had none at hand, and therefore set myself to devise some other form of verbal torture which should serve the same purpose. The result of my meditations was a new kind of Puzzle—new at least to me—which, now that it has been fairly tested by a year's experience and commended by many friends, I offer to you, as a newly-gathered nut, to be cracked by the omnivorous teeth which have already masticated so many of your Double Acrostics.

The rules of the Puzzle are simple enough. Two words are proposed, of the same length ; and the Puzzle consists in linking these together by interposing other words, each of which shall differ from the next word *in one letter only*. That is to say, one letter may be changed in one of the given words, then one letter in the word so obtained, and so on, till we arrive at the other given word. The letters must not be interchanged among themselves, but each must keep to its own place. As an example, the word "head" may be changed into "tail" by interposing the words "heal, teal, tell, tall." I call the two given words "a Doublet," the interposed words " Links," and the entire series "a Chain," of which I here append an example :—

```
H E A D
h e a l
t e a l
t e l l
t a l l
T A I L
```

It is, perhaps, needless to state that it is *de rigueur* that the links should be English words (including well-known names), such as might be used in good society.

The easiest "Doublets" are those in which the consonants in one word answer to consonants in the other, and the vowels to vowels ; "head" and "tail" are a Doublet of this kind. Where this is not the case, as in "head" and "bare," the first thing to be done is to transform one member of the Doublet into a word whose consonants and vowels shall answer to those in the other member (*e.g.*, "head, herd, here"), after which there is seldom much difficulty in completing the "Chain."

I am told that there is an American game involving a similar principle. I have never seen it, and can only say of its inventors, "*pereant qui ante nos nostra dixerunt!*"

LEWIS CARROLL.

DOUBLETS.

1. Drive PIG into STY.
2. Raise FOUR to FIVE.
3. Make WHEAT into BREAD.

No answer can be acknowledged unless it be received at "Vanity Fair" Office by twelve o'clock at noon next Thursday.

March. 24. 1879.

DOUBLETS.

THIS Puzzle consists in linking together two given words of the same length by interposing certain other words in accordance with the following

RULES.

1. The words given to be linked together constitute a "Doublet;" the interposed words are the "Links;" and the entire series a "Chain." The object is to complete the Chain with the least possible number of Links.

2. Each word in the Chain must be formed from the preceding word by changing one letter in it, and one only. The substituted letter must occupy the same place, in the word so formed, which the discarded letter occupied in the preceding word, and all the other letters must retain their places.

3. When three or more words are given to be made into a Chain, the first and last constitute the "Doublet." The others are called "Set Links," and must be introduced into the Chain in the order in which they are given. A Chain of this kind must not contain any word twice over.

4. No word is admissible as a Link unless it (or, if it be an inflection, the word from which it comes) is to be found in some known Dictionary, and is also a word which might be used, and would be universally understood, in good Society.

The following are inadmissible :—

  a. Words marked "local" in the Dictionary, and Scotticisms such as "auld" and "ain."
  b. French, Latin, and other foreign words, with the exception of those which (like "ennui," "minimum," "kudos," "loot") have been so thoroughly naturalised as to be virtually English words.
  c. Proper names.
  d. Abbreviations such as "stept" for "stepped," "e'en" for "even," "e'er" for "ever."
  e. A combination of two words which is usually printed without a hyphen (such as "teapot") is admissible as a Link; but not if (like "tea-set") it is usually printed as two words. The diphthong æ, œ, and qu are counted as single letters.

5. The marks assigned to each Doublet are as follows :—If it be given without any Set Links, so many marks are assigned to it as there are letters in the two words together (*e.g.*, a four-letter Doublet would have eight marks assigned to it). If it be given with Set Links, so that the Chain is made up of two or more portions, so many marks are assigned to it as would have been assigned if each portion had been a separate Chain (*e.g.*, a four-letter Doublet which has two Set Links, so that the Chain is made up of three portions, would have twenty-four marks assigned to it).

6. Each competitor, who completes the Chain with the least possible number of Links, will receive the full number of marks assigned; and each who uses more than the least possible number of Links will lose a mark for every additional Link.

Difficulties will, no doubt, sometimes arise in the application of Rule 4, whenever a word is used as a Link which lies close on the border-line dividing the Admissible from the Inadmissible. All such " Hard Cases " will be settled by the exercise of a dictatorial authority on the part of CHOKER, *from whose decision there is no appeal.* Any competitor who feels doubtful as to the admissibility of any word which he has used as a Link is recommended to send in a second Chain. not containing the doubtful word ; and if more than one of the Chains thus sent in are found to be admissible, CHOKER will give him credit for the shortest of them.

LEWIS CARROLL.

The Competition for Doublet Prizes begins with the Doublets given in the present number, and will include all the Doublets published up to and on the 26th July.

Three Prizes, consisting for the first of a Proof Album, and for the second and third of ordinary Albums, will be given to the three Competitors who during this period score the largest number of marks.

No answer to the Doublets can be acknowledged unless it be written on a separate piece of paper, and be received at "Vanity "Fair" Office, 12, Tavistock Street, Covent Garden, by 12 at noon on the Saturday following the date of the number in which the Doublets have been given.

The answers to the Doublets will be published, together with the score made by each competitor, in the number of "Vanity Fair" published next after the latest day for receiving the answers.

NEW METHOD OF SCORING.

D EAR VANITY,—The commencement of the second year of the Doublets competition seems to afford a good opportunity for introducing a change into the system of marking. I have considered the subject very carefully, and have come to the conclusion that the present system does not accurately measure the skill employed, and that I can suggest one which will do more justice to the rival merits of your little army of Doubleteers.

I propose, then, to substitute for Rules 1, 2, in the chapter headed "Method of Scoring," the following :—

1. The shortest Chain which can be made on a given Doublet will have so many marks assigned to it as there are letters in the Links employed, "Set Links" counting as ordinary Links.

2. Each competitor who completes his Chain with the least possible number of Links will receive the marks assigned by Rule 1 ; and each who uses more than the least possible number will forfeit, for every extra Link, as many marks as there are letters in it.

To illustrate the New Rules, let us take the Doublet (the first one ever published) " Drive PIG into STY." The shortest known Chain for this ("PIG, wig, wag, way, say, STY") contains 4 Links. Here a competitor using only 4 Links would score 12 marks ; one using 5 Links, 9 marks ; 6 Links, 6 marks ; 7 Links, 3 marks : and any competitor using 8 or more Links would score nothing.

The points of agreement and of difference between the two systems will be best illustrated by examples.

Take a 3-Letter Doublet and a 4-Letter Doublet, and suppose that the shortest Chains made on them contain 5 Links each. By the present system one would score 6, the other 8 ; by the new, one would score 15, the other 20. Here, so far as the proportion is concerned, the two systems agree.

Again, take two 3-letter Doublets, and suppose that the shortest Chains made on them contain, respectively, 4 Links and 8 Links. By the present system each would score 6 ; by the new, one would score 12, the other 24. This is surely more just, since the second would require about twice as much mental labour as the first.

Again, take a 3-letter Doublet and a 6-letter Doublet, and suppose there are two competitors, one of whom beats the other, on each Doublet, by one Link. By the present system he would gain one mark in each case ; by the new, he would gain 3 marks in the first, and 6 in the second. And surely this also is more just, since it would require about twice as much mental labour to save a 6-letter Link as to save a 3-letter one.

I feel confident that your adoption of the new system will prove satisfactory to your readers, and that the future drivers of (mental) Pigs into (mental) Sties will find their skill more exactly measured, and therefore more justly rewarded.

LEWIS CARROLL.

DOUBLETS ALREADY SET.

March 29.—Drive PIG into STY.
Raise FOUR to FIVE.
Make WHEAT into BREAD.

April 5.—Dip PEN into INK.
Touch CHIN with NOSE.
Change TEARS into SMILE.

April 12.—Change WET to DRY.
Make HARE into SOUP.
PITCH TENTS.

April 19.—Cover EYE with LID.
Prove PITY to be GOOD.
STEAL COINS.

April 26.—Make EEL into PIE.
Turn POOR into RICH.
Prove RAVEN to be MISER.

May 3.—Change OAT to RYE.
Get WOOD from TREE.
Prove GRASS to be GREEN.

The next Competition will commence with the Doublets set on the 2nd of August.

Before that date it is intended to publish a Glossary of Admissible Words, which will of course supersede Rule 4.

May 3, 1879.

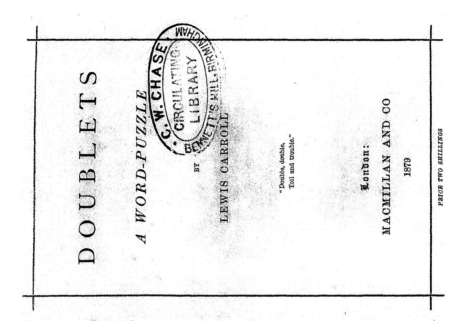

DOUBLETS

A WORD-PUZZLE

BY

LEWIS CARROLL

"Double, double,
Toil and trouble."

London:
MACMILLAN AND CO
1879

PRICE TWO SHILLINGS

3

PREFACE.

On the 29th of March, 1879, the following article appeared in "VANITY FAIR:"—

A NEW PUZZLE.

The readers of *Vanity Fair* have during the last ten years shown so much interest in the Acrostics and Hard Cases which were first made the object of sustained competition for prizes in this journal, that it has been sought to invent for them an entirely new kind of Puzzle, such as would interest them equally with those that have already been so successful. The subjoined letter from Mr. Lewis Carroll will explain itself, and will introduce a Puzzle so entirely novel and withal so interesting, that the transmutation of the original into the final word of the Doublets may be expected to become an occupation to the full as amusing as the guessing of the Double Acrostics has already proved.

In order to enable readers to become acquainted with the new Puzzle, preliminary Doublets will be given during the next three weeks—that is to say, in the present number of *Vanity Fair* and in those of the 5th and 12th April. A competition will then be opened—beginning with the Doublets published on the 19th April, and including all those published subsequently up to and including the number of the 26th July—for three prizes, consisting respectively of a Proof Album for the first and of Ordinary Albums for the second and third prizes.

The rule of scoring will be as follows:—A number of marks will be apportioned to each Doublet equal to the number of letters in the two

A 2

𝕴𝖓𝖘𝖈𝖗𝖎𝖇𝖊𝖉

to

𝕵𝖚𝖑𝖎𝖆 𝖆𝖓𝖉 𝕰𝖙𝖍𝖊𝖑

the younger Arnolds — sisters to Mrs Humphry Ward

PREFACE.

"a Doublet," the interposed words "Links," and the entire series "a Chain," of which I here append an example:—

```
H E A D
h e a l
t e a l
t e l l
t a l l
T A I L
```

It is, perhaps, needless to state that it is *de rigueur* that the links should be English words, such as might be used in good society.

The easiest "Doublets" are those in which the consonants in one word answer to consonants in the other, and the vowels to vowels; "head" and "tail" constitute a Doublet of this kind. Where this is not the case, as in "head" and "hare," the first thing to be done is to transform one member of the Doublet into a word whose consonants and vowels shall answer to those in the other member (e.g., "head, herd, here"), after which there is seldom much difficulty in completing the "Chain."

I am told that there is an American game involving a similar principle. I have never seen it, and can only say of its inventors, "*pereant qui ante nos nostra dixerunt!*"

LEWIS CARROLL.

PREFACE.

words given. For example, in the instance given below of "Head" and "Tail," the number of possible marks to be gained would be eight; and this maximum will be gained by each one of those who make the chain with the least possible number of changes. If it be assumed that in this instance the chain cannot be completed with less than the four links given, then those who complete it with four links only will receive eight marks, while a mark will be deducted for every extra link used beyond four. Any competitor, therefore, using five links would score seven marks, any competitor using eight links would score four, and any using twelve links or more would score nothing. The marks gained by each competitor will be published each week.

DEAR VANITY,—Just a year ago last Christmas, two young ladies—smarting under that sorest scourge of feminine humanity, the having "nothing to do"—besought me to send them "some riddles." But riddles I had none at hand, and therefore set myself to devise some other form of verbal torture which should serve the same purpose. The result of my meditations was a new kind of Puzzle—new at least to me—which, now that it has been fairly tested by a year's experience and commended by many friends, I offer to you, as a newly-gathered nut, to be cracked by the omnivorous teeth which have already masticated so many of your Double Acrostics.

The rules of the Puzzle are simple enough. Two words are proposed, of the same length; and the Puzzle consists in linking these together by interposing other words, each of which shall differ from the next word *in one letter only*. That is to say, one letter may be changed in one of the given words, then one letter in the word so obtained, and so on, till we arrive at the other given word. The letters must not be interchanged among themselves, but each must keep to its own place. As an example, the word "head" may be changed into "tail" by interposing the words "heal, teal, tell, tall." I call the two given words

RULES.

1. The words given to be linked together constitute a "Doublet;" the interposed words are the "Links;" and the entire series a "Chain." The object is to complete the Chain with the least possible number of Links.

2. Each word in the Chain must be formed from the preceding word by changing one letter in it, and one only. The substituted letter must occupy the same place, in the word so formed, which the discarded letter occupied in the preceding word, and all the other letters must retain their places.

3. When three or more words are given to be made into a Chain, the first and last constitute the "Doublet." The others are called "Set Links," and must be introduced into the Chain in the order in which they are given. A Chain of this kind must not contain any word twice over.

4. No word is admissible as a Link unless it (or, if it be an inflection, a word from which it comes) is to be found in the following Glossary. Comparatives and superlatives of adjectives and adverbs, when regularly formed, are regarded as 'inflections' of the positive form, and are not given separately: e.g. the word 'new' being given, it is to be understood that 'newer' and 'newest' are also admissible. But nouns formed from verbs (as 'reader' from 'read') are *not* so regarded, and may not be used as Links unless they are to be found in the Glossary.

METHOD OF SCORING, &c.

ADOPTED IN "VANITY FAIR."

1. The marks assigned to each Doublet are as follows:—If it be given without any Set Links, so many marks are assigned to it as there are letters in the two words together (e.g., a four-letter Doublet would have eight marks assigned to it). If it be given with Set Links, so that the Chain is made up of two or more portions, so many marks are assigned to it as would have been assigned if each portion had been a separate Chain (e.g., a four-letter Doublet which has two Set Links, so that the Chain is made up of three portions, would have twenty-four marks assigned to it).

2. Each competitor, who completes the Chain with the least possible number of Links, will receive the full number of marks assigned; and each who uses more than the least possible number of Links will lose a mark for every additional Link.

3. Each competitor is required to send his three Chains, with his signature attached, written on *one* piece of paper.

4. The Editor of 'Vanity Fair' will be glad to receive any suggestions, both as to words which it seems desirable to omit, and as to omitted words which it seems desirable to insert: but any word proposed for insertion or for omission *should be exhibited as a Link between two other words.*

5. Alterations will not be made in this Glossary during any competition, but will be duly announced before the commencement of a new competition, so that those who already possess copies will be able to correct them, and will not be obliged to buy a new edition.

"Vanity Fair" Office,
13, Tavistock Street,
Covent Garden,
LONDON.

[106]

DOUBLETS ALREADY SET

IN "VANITY FAIR."

March 29.—Drive PIG into STY.
Raise FOUR to FIVE.
Make WHEAT into BREAD.

April 5.—Dip PEN into INK.
Touch CHIN with NOSE.
Change TEARS into SMILE.

April 12.—Change WET to DRY.
Make HARE into SOUP.
PITCH TENTS.

April 19.—Cover EYE with LID.
Prove PITY to be GOOD.
STEAL COINS.

April 26.—Make EEL into PIE.
Turn POOR into RICH.
Prove RAVEN to be MISER.

May 3.—Change OAT to RYE.
Get WOOD from TREE.
Prove GRASS to be GREEN.

May 10.—Evolve MAN from APE.
Change CAIN into ABEL.
Make FLOUR into BREAD.

DOUBLETS ALREADY SET.

May 17.—Make TEA HOT.
Run COMB into HAIR.
Prove a ROGUE to be a BEAST.

May 24.—Change ELM into OAK.
Combine ARMY and NAVY.
Place BEANS on SHELF.

May 31.—HOOK FISH.
QUELL a BRAVO.
Stow FURIES in BARREL.

June 7.—BUY an ASS.
Get COAL from MINE.
Pay COSTS in PENCE.

June 14.—Raise ONE to TWO.
Change BLUE to PINK.
Change BLACK to WHITE.

June 21.—Change FISH to BIRD.
Sell SHOES for CRUST.
Make KETTLE HOLDER.

N.B. Solutions of these Doublets will be found at *p.* 38.

8

PREFACE TO GLOSSARY.

The following Glossary is intended to contain all well-known English words (or, if they are inflections, words from which they come) of 3, 4, 5, or 6 letters each, which may be used in good Society, and which can serve as Links. It is not intended to be used as a source from which words may be obtained, but only as a test of their being admissible.

That such a Glossary is needed may best be proved by quoting the following passage from 'Vanity Fair' of May 17, 1870, premising that all the strange words here used, had actually occurred in Chains sent in by competitors:—

"CHOKER humbly presents his compliments to the four thousand three hundred and seventeen (or thereabouts) indignant Doubleteers who have so strongly shent him, and pre to being stoaked in the spate of their wrath, asks for a fiver of minutes for reflection. CHOKER is in a state of complete pye. He feels that there must be a stent to the admission of spick words. He is quite unable to sweal the chaffy spelt, to sile the pory cole, or to swill's spate from a piny ait to the song of the spink. Frils and the mystic Gole are strangers in his sheal: the chancetul Gord hath never brought him gold, nor ever did a cate become his sin. The Doubleteers will no doubt spank him sore, with slick quotations and wild words of yore, will pour upon his head whole steres of steens and poods of spiles points downwards. But he trusts that those alone who habitually use such words as these in Good Society,

and whose discourse is universally there understood, will be the first to cast a steam at him."

As the chief object aimed at has been to furnish a puzzle which shall be an amusing mental occupation at *all* times, whether a dictionary is at hand or not, it has been sought to include in this Glossary only such words as most educated people carry in their memories. If any doubt should arise as to whether any word that suggests itself is an admissible one, it may be settled by referring to the Glossary.

When there are two words spelt alike, one a noun and one a verb, or any other such combination, it has not been thought necessary to include *both*, so long as all the inflections can be obtained from *one*: e.g. 'aim' is given only as a verb, since 'aims,' the plural of the noun, is also the third person of the verb; but 'hale, v.a.,' and 'hale, a.,' are both given, the one being needed to supply 'hales' and 'haled,' and the other to supply 'haler.'

Two abbreviations, 'e'en' and 'e'er,' have been included.

As to the many words which, though used and understood in good Society, are yet not available as Links, owing to there being no other words into which they can be changed, it has been regarded as a matter of indifference whether they are included or not.

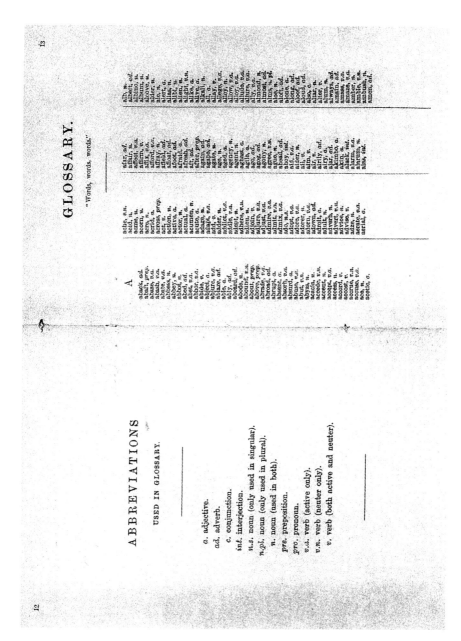

ABBREVIATIONS

USED IN GLOSSARY.

a. adjective.

ad. adverb.

c. conjunction.

int. interjection.

n.s. noun (only used in singular).

n.pl. noun (only used in plural).

n. noun (used in both).

pre. preposition.

pro. pronoun.

v.a. verb (active only).

v.n. verb (neuter only).

v. verb (both active and neuter).

GLOSSARY.

"Words, words, words."

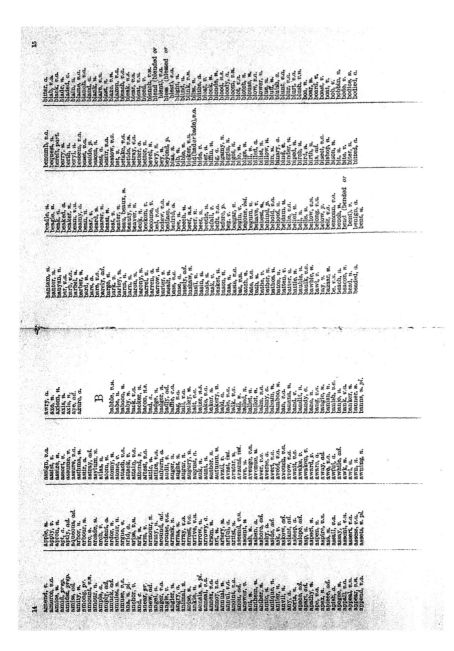

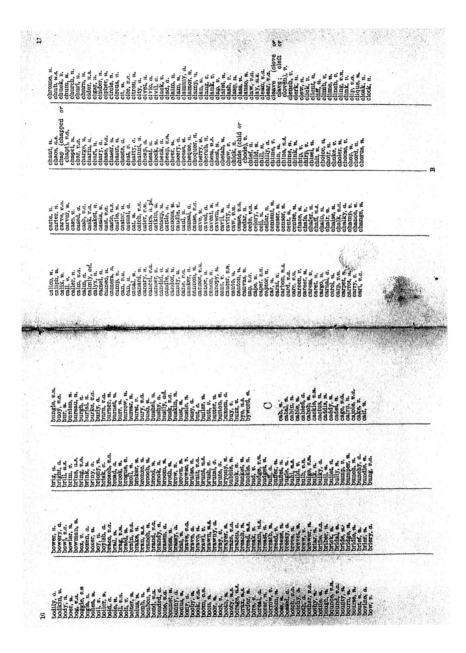

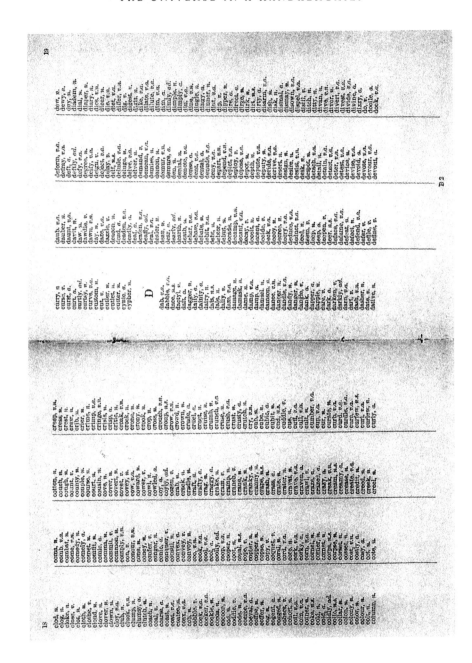

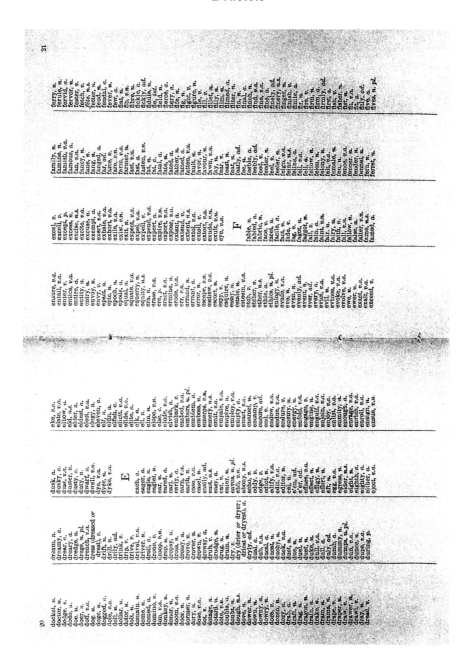

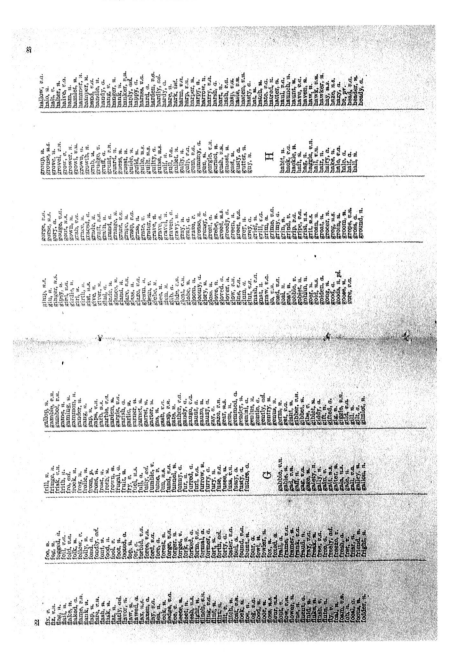

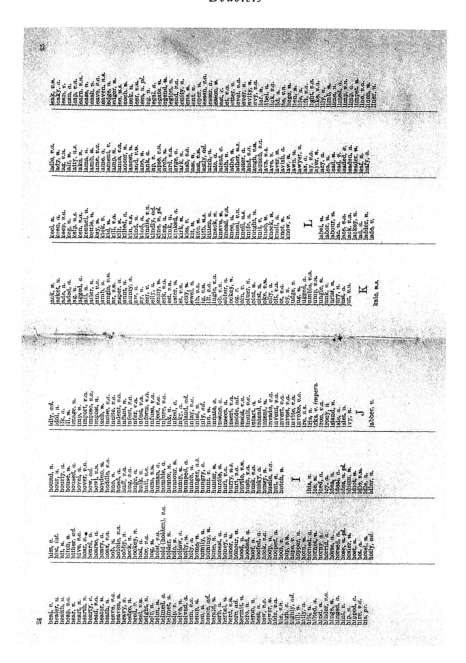

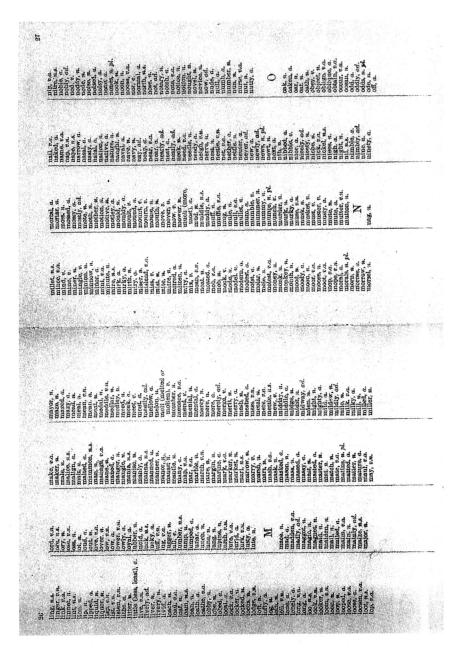

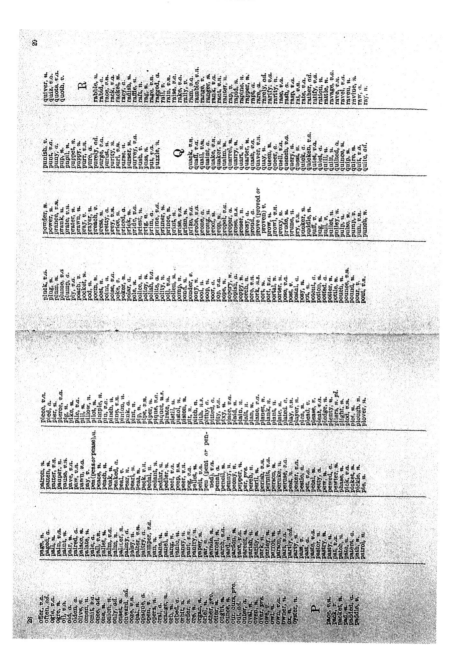

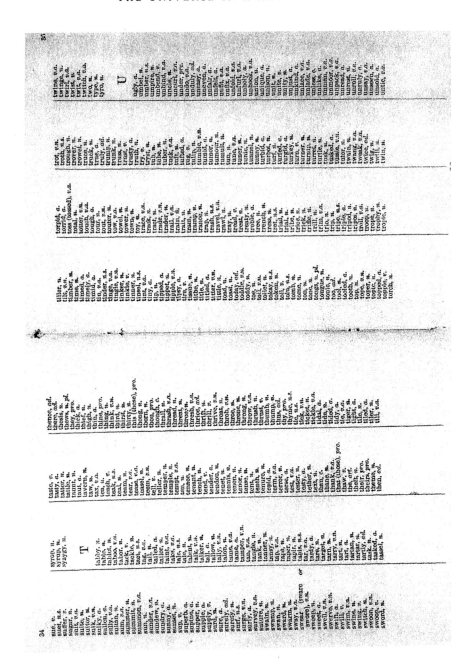

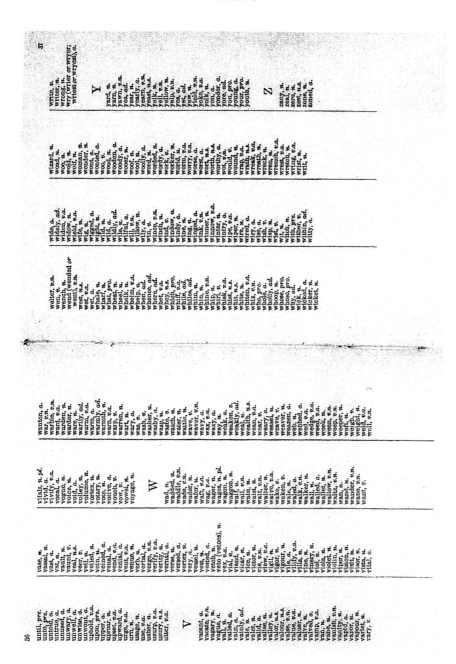

write, v.
writer, n.
wrong, n.
wry (wrier or wryer;
 wriest or wryest), a.

Y

yard, n.
yarn, n.
yawn, v.n.
yea, ad.
year, n.
yearly, a.
yearn, v.n.
yeast, n.s.
yell, v.n.
yellow, n.
yelp, v.n.
yes, c.
yes, ad.
yet, c.
yield, v.n.
yoke, v.a.
yolk, n.
you, pl.
yonder, a.
yore, ad.
you, pro.
young, a.
your, pro.
youth, n.

Z

zany, n.
zeal, n.
zero, n.s.
zest, n.s.
zone, n.
zoned, a.

---

wizard, n.
witch, n.
woe, n.
wold, n.
wolf, n.
woman, n.
wonder, v.n.
wont, a.
wonted, a.
woo, v.
wood, n.
wooden, a.
woody, a.
wooer, n.
woof, n.
wool, n.
woolly, a.
word, n.
worded, a.
wordy, a.
work, v.
worker, n.
world, n.
worm, v.a.
worry, v.a.
worse, a.
worst, v.
wort, n.s.
worth, n.s.
worthy, a.
would, v.
wound, n.
wrap, v.a.
wrath, n.s.
wreak, v.a.
wreath, n.
wreck, n.
wren, n.
wrench, v.a.
wrest, v.a.
wring, v.a.
wrist, n.
writ, n.

---

wide, a. ad.
widely, ad.
widen, v.a.
widow, n.
wield, v.a.
wife, n.
wig, n.
wigged, a.
wight, n.
wild, a.
wild, n.
wildly, ad.
wile, v.
wilful, a.
will, v.n.
willow, n.
wily, a.
win, v.
wince, v.n.
winch, n.
wind, v.
window, n.
windy, a.
wine, n.
wing, v.a.
winged, a.
wink, v.n.
winner, n.
winnow, v.a.
winter, n.
wintry, a.
wiper, n.
wire, a.
wired, a.
wiry, a.
wish, v.
wisp, n.
wist, v.
witch, v.
with, pre.
wither, ad.
within, ad.
witty, a.

---

welter, v.n.
wen, n.
wend (wended or
 went), v.a.
west, n.s.
wet, v.a.
whale, v.
wharf, n.
what, pro.
wheat, n.
wheel, v.
whelk, n.
whelm, v.a.
whelp, n.
when, ad.
whence, ad.
where, ad.
whet, v.a.
whey, n.
which, pro.
whiff, v.a.
while, ad.
whim, n.
whine, v.n.
whip, n.
whir, n.
whisk, v.a.
whist, n.s.
white, a.
whiten, v.a.
whit, n.s.
who, pro.
whole, a.
wholly, ad.
whoop, n.
whoop, pro.
whose, pro.
why, ad.
wicker, n.
wicked, a.
wicket, n.

---

until, pre.
unto, pre.
untrue, a.
unused, a.
unwary, a.
unwell, a.
unwise, a.
uphold, v.a.
upon, pre.
upper, a.
uproar, n.
upset, v.a.
upward, a.
urge, v.a.
urn, n.
usage, n.
use, v.a.
usher, n.
usury, v.a.
utter, v.a.

V

vacant, a.
vacate, v.a.
vagary, n.
vail, v.
vailed, a.
vain, a.
vainly, ad.
vale, n.
valet, n.
valid, a.
valley, n.
valor, n.s.
valour, n.s.
value, a.
valuer, n.
valve, v.
valved, a.
vamp, v.a.
vane, n.
vanish, v.n.
vanity, n.
vapid, a.
vapor, n.
vapour, n.
varlet, n.
vary, v.

---

vase, n.
vassal, n.
vast, a.
vat, n.
vault, v.
veal, n.s.
veer, v.
veil, n.
veiled, a.
vein, n.
veined, a.
vend, v.a.
venial, a.
vent, v.a.
venue, n.
verb, n.
verbal, a.
verge, v.a.
verify, v.a.
verily, a.
vernal, a.
verse, a.
versed, a.
vertex, n.
vessel, n.
vest, n.
vested, a.
veto (vetoes), n.
vetch, a.
vial, v.a.
vial, n.
viand, n.
vicar, n.
vice, n.
vigor, n.
vigour, n.
vigil, n.
view, v.a.
vigour, n.
vile, a.
vilify, v.a.
villa, n.
vine, n.
vinery, n.
viol, n.
violet, n.
violin, n.
viper, n.
virus, n.
visor, n.
vista, n.
vital, a.

---

vitals, n. pl.
vivid, a.
vividly, v.a.
vocal, a.
vogue, n.
voice, n.
void, a.
volley, n.
volume, n.
vortex, n.
votary, n.
vote, n.
votive, a.
vouch, v.
vow, v.
vowel, n.
voyage, n.

W

wad, n.
wadded, a.
waddle, v.n.
wade, v.n.
wafer, n.
waft, v.a.
wag, v.a.
wager, n.
wagon, n. pl.
waggon, n.
waif, n.
wail, v.
wain, n.
waist, n.
waiter, n.
wake, v.
waken, v.
walk, v.a.
walker, n.
wall, n.
walled, a.
wallet, n.
wallow, v.n.
waltz, v.n.
wan, a.
wander, v.n.
wane, v.n.
want, v.

---

wanton, a.
war, v.n.
warble, v.n.
warden, n.
warder, n.
ware, n.
warily, ad.
warm, v.a.
warmly, ad.
warmth, n.
warn, v.a.
warp, v.
warren, n.
wart, n.
wary, a.
was, v.
wash, v.
washer, n.
washy, a.
wasp, n.
waste, v.
watch, n.
water, n.
wattle, v.a.
wave, v.n.
wavy, a.
wax, n.
waxy, a.
way, n.
weak, a.
weaken, v.
weakly, ad.
wealth, n.s.
wean, v.
weary, a.
weasel, n.
weave, v.
weaver, n.
weazen, a.
webbed, a.
wedge, v.a.
weed, v.a.
week, n.
weekly, v.n.
weep, v.n.
weeper, n.
weft, n.
weigh, v.
weld, v.a.
well, v.a.
well, v.n.

39

SHOES
shoes
floss
gloss
cross
cress
CRUST

FURIES
buries
buried
barked
barbed
BARREL

KETTLE
settle
settee
betted
belted
bolted
boulder
HOLDER

QUELL
quill
quilt
guilt
guild
grade
grave
BRAVO

COSTS
posts
pests
tests
tents
pence
PENCE

BLACK
clack
crack
track
trick
trice
white
WHITE

GRASS
cross
cress
tress
freed
greed
GREEN

FLOUR
floor
flood
blood
brood
broad
BREAD

ROGUE
vogue
vague
value
valve
heave
heart
least
BEAST

BEANS
beams
seams
shams
shame
shale
shell
SHELF

WHEAT
cheat
cheap
cheep
creep
creed
breed
BREAD

TEARS
sears
stars
stare
stale
stile
SMILE

PITCH
pinch
winch
wench
tench
tents
TENTS

STEAL
steel
steer
sheer
shier
shins
coins
COINS

RAVEN
riven
rivet
river
riser
MISER

SOLUTIONS OF DOUBLETS. (See p. 8.)

MINE
mint
mist
most
moat
coal
COAL

BLUE
flue
glue
glut
gout
pout
punt
pint
PINK

FISH
wish
wash
want
ward
bard
BIRD

POOR
boor
book
rook
rock
rich
RICH

TREE
free
flee
feed
weed
wend
wild
wood
WOOD

COMB
come
core
bore
bare
hair
HAIR

ARMY
arms
aims
dims
dams
dame
name
navy
NAVY

HOOK
book
boot
bolt
boat
brat
bran
bird
FISH

FOUR
foul
fool
foot
fort
fore
fire
FIVE

NOSE
note
core
cone
conn
chin
CHIN

HARE
bare
bore
bord
word
wood
soud
SOUP

PITY
pits
pins
pint
find
fond
food
good
GOOD

CAIN
chin
shin
spin
sped
sped
abed
ABEL

OAT
rat
rot
rye
RYE

TEA
sea
set
sot
hot
HOT

ELM
ell
all
ail
air
oar
OAK

BUY
bud
bid
aid
and
ask
ASS

ONE
ore
ere
err
ear
tar
too
TWO

38

PIG
pit
wit
wat
way
say
STY

PEN
een
ten
tin
win
ink
INK

WET
bet
bey
dey
dry
DRY

EYE
dye
die
did
lid
LID

EEL
een
pen
pie
PIE

APE
are
ore
ere
err
mar
MAN

7

Pamphlets on Games

☞ Court Circular ☜

Carroll records in his diary (January 28, 1858): "Completed the rules of the game of cards I have been inventing during the past few days, *Court Circular*." His anonymous three-page pamphlet on the game, reprinted below, was issued in 1860. It was reprinted anonymously two years later, with simplified rules. Here is the 1860 version:

RULES

FOR

COURT CIRCULAR.

(A New Game of Cards for Two or More Players.)

SECTION I. *(For Two Players.)*

I.

Cut for precedence. Highest is "first-hand;" lowest "dealer." Dealer gives 6 cards to each, one by one, beginning with first-hand, and turns up the 13th, which is called the "Lead." It is convenient that the same player should be dealer for the whole of each game.

II.

First-hand then plays a card; then the other player, and so on, until 6 cards have been played, when the trick is complete, and he who can make, (out of the 3 cards he has played, with or without the Lead), the best "Line," wins it.

First-hand.

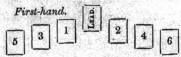

N.B. The cards in the figure are numbered in the order of playing.

III.

A "Line" consists of 2, or all 3, of the cards put down by either player, with or without the Lead. In making a Line, it does not matter in what order the 3 cards have been put down. Lines rank as follows :

(1) 3, OR 4 CARDS, (LEAD *included*.)

Trio—i.e. 3 of a sort, (e.g. 3 Kings, or 3 Nines.)

Sequence—i.e. 3, or 4, in Sequence, (e.g. Eight, Nine, Ten, Knave.)

Sympathy—i.e. 3, or 4, Hearts.

Court—i.e. 3, or 4, Court-cards, (if 4, it is called **Court Circular.**)

N.B. In this Class a Line of 4 cards beats a *similar* Line of 3. The Lead must not be reckoned in the middle of a Sequence.

(2) 3 CARDS, (LEAD *excluded*.)

Names as above.

N.B. In making a Sequence, the Ace may be reckoned either with King, Queen, or with Two, Three.

(3) 2 CARDS, (LEAD *excluded*.)

Pair—i.e. 2 of a sort.

Valentine—i.e. 2 Hearts.

Etiquette—i.e. 2 Court-cards.

IV.

If both have made Lines of the same kind, he whose Line contains the best card wins the trick ; and if neither has made a Line, he who has played the best card wins it. Cards rank as follows :

(1) Hearts.

(2) The rest of the pack, in the order Aces, Kings, &c.

N.B. If no Hearts have been played, and the highest cards on each side are equal, (e.g. if each have played an Ace,) they rank in the order Diamonds, Clubs, Spades.

V.

The winner of a trick chooses, as Lead for the next trick, any one of the cards on the table, except the old Lead ; he then takes the rest, turning them face upwards, if he be first-hand, but if not, face downwards ; and he becomes first-hand for the next trick.

VI.

The dealer then gives cards to each, one by one, beginning with first-hand, until each hand is made up again to 6 cards.

VII.

At any time during a trick, after the first card of it has been played, and before either has played 3 cards, he whose turn it is to play may "resign" instead.; in which case no more cards are played in that trick, and the other player wins it and proceeds as in Rule V. But when either has played 3 cards, the other must not resign.

VIII.

When the pack is exhausted neither player may resign. The winner of the last trick clears the board. Each then reckons up the cards he has won, which count as follows :

Cards face upwards - - - - 2 each.
downwards - - - 1
Hearts - - - - - - 1
Court-cards - - - - - 1

(so that a Court-Heart, if face upwards, counts 4 altogether.) The winner scores the difference between his own and the loser's marks, the loser scoring nothing. Game is 20 or 50.

SECTION II. (*For Three or More Players.*)

The same rules apply, with the following necessary changes. The Lead is placed in the middle ; first-hand then plays a card ; then the player on his left-hand, and so on all round, each putting down his 3 cards in a row from the Lead towards himself. He who makes the best Line wins the trick, and is first-hand for the next trick. At any time during a trick, after the first card of it has been played, and before any one has played 3 cards, he whose turn it is to play may "resign" instead ; in which case he loses his chance of winning that trick, and the other players go on without him. But when any one has played 3 cards, no other player may resign. In the case where all the players but one "resign," he who is left to the last wins the trick. At the end of each game all the players but the lowest score the difference between their own marks and those of the lowest, the lowest scoring nothing. Game is 50.

January, 1860.

Croquet Castles

"Wrote out the rules of a new croquet game, for five players," Carroll wrote in his diary (May 4, 1863), "which I have invented and think of calling *Croquet Castles*." He invented the game, and kept improving it, while playing croquet with Alice Liddell and her two sisters. *Croquet Castles: For Five Players,* Carroll's first printed version of the game, was an anonymous three-page pamphlet printed that same year. In 1866 he expanded it to four pages and revised it under the new title *Castle-Croquet for Four Players.* The rules were reprinted in *Aunt Judy's Magazine* (August 1867). The rules given in Collingwood's *Lewis Carroll Picture Book* differ slightly from both of the two pamphlets reprinted here, and may have been based on a third or fourth printing.

CROQUÊT CASTLES.

FOR FIVE PLAYERS.

I.

THIS Game requires the 10 arches, and 5 of the 8 balls used in the ordinary game, and, in addition to them, another set of 5 balls, (matching these in colour, but marked so as to be distinct from them), and 5 flags, also matching them. One set of balls is called "soldiers"; the other, "sentinels." The arches and flags are set up as in the figure, making 5 "castles," and each player has a castle, a soldier, and a sentinel ; the sentinel's "post" is half-way between the "gate" and the "door" of the castle, and the soldier is placed, to begin the game, just within the gate.

(N.B.—The distance from one gate to the next should be 6 or 8 yards, and from the gate of a castle to the door 4 yards ; and the distance from the door to the flag should be equal to the width of the door.)

II.

The soldiers are played in order, as marked above ; then the sentinels, in the same order, and so on. Each soldier has to "invade" the other 4 castles, in order, (e. g. soldier No. 3 has to invade castles Nos. 4, 5, 1, 2,) then to re-enter his own, and touch the flag ; and whoever does this first, wins. To "invade" a castle, he must enter the gate, go through the door, then between the door and the flag, then out at the gate again : but he cannot enter a castle, unless either the sentinel of that castle, or his own sentinel, be out of its castle.

(N.B. No ball can enter or leave a castle, except by passing through the gate.)

III.

If a sentinel touch a soldier, both being in the sentinel's castle, the soldier is "prisoner ;" he is replaced (if necessary) where he was when touched, the sentinel is placed in the gate, and the castle is "fortified." The prisoner cannot move, and nothing can go through the gate, till the castle is opened again, which is done either by the prisoner's comrade coming and touching the sentinel in the gate, or by the sentinel leaving the gate to go and rescue his own comrade : in the former case, both sentinels are replaced at their posts.

IV.

When a prisoner is set free, he cannot be again taken prisoner until after his next turn.

V.

If a ball touch another, (except a prisoner, or a sentinel in his castle,) the player may, if he likes, replace it where it was when touched, and use it to croquêt his own with : in the excepted cases, he must replace it, but can do no more.

VI.

If a soldier go through an arch, or between a door and flag, in his proper course, or if a sentinel go through the gate of his castle, the player has another turn.

VII.

A player, whose soldier is a prisoner, plays all his turns with his sentinel; and one, whose castle is fortified, with his soldier, unless it be taken prisoner, when he must play his sentinel to rescue it.

VIII.

The sentinel of a fortified castle is considered to be in, or out of, the castle, as the owner chooses: that is, if he wishes to invade a castle, the sentinel of which is within it, he may consider his own sentinel as *out of* its castle, (which gives him the right of invasion): or, if he wishes to go and rescue his soldier, he may consider it as *in*, (so that he first plays it *through* the gate, and then has another turn).

CH. CH., OXFORD, *May* 4, 1863.

N.B. This game does not absolutely require more than *two* additional balls, beside those used in the ordinary game; these may be Light Blue and Light Green, and the 10 balls may be arranged as follows:—

Soldiers.	*Sentinels.*
BLUE.	LIGHT BLUE.
BLACK.	BROWN.
ORANGE.	YELLOW.
GREEN.	LIGHT GREEN.
RED.	PINK.

CASTLE-CROQUÊT.
FOR FOUR PLAYERS.

```
            .
            .
            .
           ..
           3
  .   .. : 2   4 :  ..   .
           1
          .. gate

          : door

          . peg
```

RULES.

I.

This game requires 8 balls, 8 arches, and 4 pegs : 4 of the balls are called " soldiers ;" the others, "sentinels." The arches and pegs are set up as in the figure, making 4 " castles," and each player has a castle, a soldier, and a sentinel. Before the game begins, each player places his sentinel within a mallet's length of his peg, and does the same with his soldier when his turn comes to play.

(N.B. The distance from one gate to the next should be 6 or 8 yards, and the distance from the gate to the door, or from the door to the peg, 2 or 3 yards.)

II.

If a sentinel goes through the gate of his castle, in the direction *from* his peg, he is said to " leave " the castle : when next he goes through it in the opposite direction, he is said to " re-enter " it, and so on. A sentinel, that has not left his castle, is said to be "on duty:" if he leaves it, he is said to be "off duty;" if he re-enters it, to be "on duty" again, and so on.

III.

To begin the game, the owner of castle No. 1 places and plays his soldier, and then plays his sentinel: then the owner of Castle No. 2, and so on. Each player has to bring his soldier out of his castle, (by playing it through the gate,) and with it "invade" the other castles in order, (*e.g.* No. 3 has to invade castles 4, 1, 2), re-enter his own castle, and lastly, touch his peg, his sentinel being "on duty" at the time; and whoever does all this first, wins. To "invade" a castle, the soldier must enter at the gate, go through the door (either way), touch the peg, and go out at the gate again.

IV.

If an invading soldier touch, or be touched by, the sentinel "on duty" of the castle he is invading, he becomes "prisoner," and is placed behind the peg. He may be released by the sentinel going "off duty," or by his own sentinel "on duty" coming and touching the peg: in the latter case, his sentinel is at once re-placed as at the beginning of the game. The released soldier is "in hand" till his next turn, when he is placed as at the beginning of the game.

V.

When a soldier goes through an arch, or touches a peg, "in order," or when a sentinel takes a prisoner, he may be played again. Also when a sentinel leaves, or re-enters, his castle, he may be played again, but may not exercise either of these privileges twice in one turn.

VI.

If the ball played touch another, (neither of them being a sentinel "on duty"), the player may "take two" off the ball so touched, but must not move it in doing so. If however the ball so touched be his own sentinel "off duty," he may take a croquêt of any kind, as in the ordinary game. He may not "take two," or take a croquêt, twice in one turn off the same ball, unless he has meanwhile gone through an arch, or touched a peg, "in order."

N.B. The following arrangement of the 8 balls as soldiers and sentinels will be found convenient :—

Castle.	Soldier.	Sentinel.
I.	Blue	Pink
II.	Black	Yellow
III.	Brown	Orange
IV.	Green	Red

ADVICE TO THE PLAYER.

As it is not easy, in a new game, to see at once what is the best method of play in the various situations that may occur, the following suggestions may be of use to the player.

There are two distinct methods of play, which you may adopt in this game, and each has its own special advantages : the one consists in keeping your sentinel " on duty ;" the other, in bringing it " off duty."

In the first method, your sentinel remains constantly at home, except when your soldier is in danger of being taken prisoner, when it is played up to the peg of the castle you are invading, so as to be ready to release your soldier. In this method, the best position for your sentinel is opposite to the centre of your gate, and a ball's width from it, so that if a soldier, trying to invade your castle, should touch it, it must have previously passed through the gate. From this position it is easy to take a prisoner in any part of your castle by the following rule :—play your sentinel just through the gate ; this gives you another turn, in which you play it in again, getting as near as possible to the invading soldier ; this gives you another turn, in which you may take it prisoner. The same process may be employed for playing your sentinel up to the peg of the castle you are invading, if it should happen that you cannot play it straight for the peg. This process, however, must not be employed when you have a prisoner in your castle, as it would be released by your sentinel going out.

In the second method, your sentinel keeps with your soldier : when playing your soldier, you carry the sentinel along with it, through one or more arches, by taking " loose croquets" or " split strokes ;" and when your soldier can do no more, you either play your sentinel close up to it, ready for the next turn, or, if your soldier is in danger of being taken prisoner, you " take two" off it, getting as close as possible to the enemy's sentinel in the first stroke, and driving it to a safe distance in the second.

The first method is the safest, when any one of the other players is better than yourself, as it enables you to prevent his entering your castle and so to delay him : but as soon as all the players, whom you have reason to fear, have passed through your castle, you had better bring your sentinel " off duty," and help on your soldier.

The second method enables you to make rapid progress in invading the other castles : you can also take prisoners almost as easily as in the first method, by " taking two" off your soldier, getting near your gate in the first stroke, and entering your castle in the second : this gives you another turn, in which you may take a prisoner. It has, however, the disadvantage of loss of time if your soldier should be made a prisoner, as in this case your sentinel has to go home, get " on duty," and return, before it can release your soldier.

If your soldier is taken prisoner, and you release it by touching the enemy's peg with your sentinel, you are in a position in which you may often retard the other players : first, by, placing your sentinel, (which is done directly after the release), in a line between your peg and an invading soldier which is aiming at it ; secondly, by placing your soldier, (which is done when your next turn comes), close to your sentinel, playing it so as to drive your sentinel in the direction of an invading soldier, and then taking it prisoner.

It evidently follows from this, that, when you have yourself taken a prisoner, and happen to be invading the castle from which it came, you should not wait till the enemy's sentinel has touched your peg and so released the prisoner, but should yourself release it, (as soon as the enemy's sentinel has nearly reached your peg), by playing your own sentinel out through your gate and in again : in this case the sentinel, which was on its way to your peg, cannot be carried back at once, but must be played all the way home.

In " taking two" off a ball you may, if you choose, play your own ball so as only just to move it, and then strike it in the direction of the other, and thus drive it to a distance. This has nearly the same effect as the " loose croquêt' of the ordinary game, but with this difference, that it does not give you the right of playing again.

If a soldier, about to invade your castle, is lying near your gate, you may take it prisoner thus :—play your sentinel out, near the soldier ; then hit it with your sentinel, and " take two" off it, so as only just to move your ball, taking care to have the soldier in a line between your sentinel and your gate ; then drive both in together ; this gives you another turn, in which you may take it prisoner.

Aug. 1866.

VINCENT, PRINTER.

⮞ Lanrick ⮜

Carroll first mentioned his chessboard game Lanrick in his diary (December 31, 1878), where he calls it "my new invention, *Natural Selection,* afterwards called *Lanrick.*" The name came from "The muster-place be Lanrick-Mead," a line in Sir Walter Scott's poem *The Lady of the Lake.*

In 1870, in a letter to May Forshall, Carroll wrote:

> Do you ever play at games? Or is your idea of life "breakfast, lessons, dinner, lessons, tea, lessons, bed, lessons, breakfast, lessons," and so on? It is a a very neat plan of life, and almost as interesting as being a sewing-machine or a coffee-grinder. (By the way, that is a very interesting question—please answer it—which would you like most to be, of those two things?) To return to the subject, if you ever *do* play games, would you see how you like my new game 'Lanrick'? I have been inventing it for about two months, and the rules have been changed almost as often as you change your mind during dinner, when you say "I'll have meat first and then pudding—no, I'll have pudding first and then meat—no, I'll have both at once—no, I'll have neither." To return to the subject, if you can think of any improvement in the Rules, please tell me. Do you know the way to improve children? *Re-*proving them is the best way.

On January 24, 1879, Carroll records playing Lanrick with two girls. On February 11, 1879, he received proofs of his single-page anonymous leaflet, "A Game for Two Players." The game's name is not on the sheet, but two months later Carroll issued an expanded set of rules in a two-page anonymous pamphlet titled *Lanrick. A Game for Two Players.* In 1880 he was back to six rules on a single leaflet with the previous title, but signed "Lewis Carroll" at the end. In 1881 a slightly reworded version, again with the same title, was issued anonymously on two pages.

The game's final version was provided in the booklet *Syzygies and Lanrick: A Word Puzzle and a Game,* published in 1893 with Lewis Carroll's name as the author. A second edition, with small changes, came out later that year. Both parts of the book are reprinted in *The Lewis Carroll Picture Book.*

LANRICK.

"The muster-place be Lanrick mead".

A GAME FOR TWO PLAYERS.

The Game is played on a chess-board, each Player having 5 men; the other requisites are a die and dice-box, and something (such as a coin) to mark a square.

The interior of the board, excluding the border-squares, is regarded as containing 6 'rows' and 6 'columns.' It must be agreed which is the first row and first column.

1.—The men *are set, alternately,* ~~Players set their~~ *in turn,* on any border-squares *they like.*

2.—The die is thrown twice, and a square marked accordingly, the first throw fixing the row, the second the column; the marked square, with the 8 surrounding squares, forms the first 'rendezvous,' into which the men are to be played.

3.—The men move like chess-queens; in playing for the first 'rendezvous,' each Player may move over 6 squares, either with one man, or dividing the move among several.

4.—When one Player has got all his men into the 'rendezvous' the other must remove from the board one of his men that has failed to get in; the die is then thrown for a new 'rendezvous,' for which each Plaper may move over as many squares as he had men in the last 'rendezvous,' and one more.

5.—If it be found that either Player has all his men already in the new 'rendezvous,' the die must be thrown again, till a 'rendezvous' is found where this is not the case.

6.—The Game ends when one Player has lost all his men.

JAN. 16, 1879.

LANRICK.

A GAME FOR TWO PLAYERS.

" The muster-place be Lanrick mead."

The game is played on a chess-board, each Player having five men. The other requisites are a die and dice-box (or a teetotum with four sides), and something (such as a coin) to mark a square.

The board is regarded as containing eight 'rows' and eight 'columns.' It must be agreed which is the first row and which is the first column.

1.—Each Player throws the die, and the one who throws highest is 'first Player.'

2.—The 'first Player' sets his men on any border-squares he likes : then the 'second Player' does the same.

3.—The die is thrown twice, all throws less than 'three' being neglected, and a square is marked accordingly, the first throw fixing the row, the second the column. The marked square forms, with the surrounding eight squares, the first 'rendezvous,' into which the men are to be played. [N.B.—Instead of neglecting all throws less than 'three,' the following rule may be adopted, If the throw be less than three, throw again ; if it be then three or more, neglect the first throw ; but if it be again less than three, double the first throw and add the second.]

4.—Each Player may move as many squares as there are men belonging to the one who has fewest, or any lesser number, either with one man, or dividing the move among several. Each man that is moved must be kept to one line, viz : either a line parallel to an edge of the board, like a rook, or a diagonal line, like a bishop.

[136]

5.—When one Player has got all his men into the 'rendezvous,' he removes from the board one of the men that have not got in, and sets the others (called 'wanderers') in the 'rendezvous.'

6.—The men in the 'rendezvous' then 'radiate,' i.e., are moved to border-squares along the eight lines which radiate from the centre of the 'rendezvous.' For all but the centre man there is no choice of direction; each must be moved along the line on which it stands : but the owner of the centre man may move it along any vacant line : if all eight be occupied, he may set it on any vacant border-square.

7.—If there were any 'wanderers,' the winner of the last 'rendezvous' may then move his men to other border-squares, moving twice as many squares as there were 'wanderers,' and not being obliged (as in Rule 4) to keep each man to one line.

8.—A new 'rendezvous' is then marked, as in Rule 3, for which the winner of the last is 'first Player.'

9.—When one Player has only two men left, the other scores as many marks as he has men : the ten men are then set again, as in rule 2, the one, who was 'second Player' when last they were set, being now 'first Player,' and the game proceeds as before.

10.—The Player, who first scores five, wins the game.

MARCH 1, 1879.

Dec. 1880

LANRICK.

A GAME FOR TWO PLAYERS.

'The muster-place be Lanrickmead.'

1. The game is played on a chess-board, each Player having five men.

2. To begin the game, one Player sets all the men on border-squares.

3. The other then selects a square set of nine squares, called a 'rendezvous,' which must not include any of his own men, and lays a mark on its centre square.

4. Both then try to get their men into this rendezvous. Each may move as many squares as he has men, or any less number, either with one man or dividing the move among several men : each man may be moved in any direction, but must, during any one turn, keep to one line of squares, whether it be straight or slanting.

5. He who did not select the rendezvous plays first. He may, instead of moving his own men, move the rendezvous-mark one square, in any direction, thus changing the position of the whole rendezvous, provided he does not move it to a border-square or so as to make the rendezvous include any of his own men ; and this he may do every turn so long as he has not moved any of his own men. When the mark is thus moved one square, any men who have got into the rendezvous must also be moved one square, so as to take the same places in the new rendezvous as they had in the one they are leaving. But whenever this would bring two men upon the same square, the mark must not be moved in that direction. This privilege, of moving the rendezvous-mark, is not allowed to the Player who laid it down.

6. When a Player has all his men in the rendezvous, he takes off the board one of those who are not in, called 'wanderers,' and moves to border-squares, in any direction, keeping each such man to one line of squares, all wanderers not already on border-squares. All other men, on both sides, keep their places, and are played from them for the next rendezvous. The other Player then selects a new rendezvous, as in Rule 3, and the game proceeds as before, until one Player has no men left.

LEWIS CARROLL.

LANRICK.

A GAME FOR TWO PLAYERS.

THIRD EDITION.

'The muster-place be Lanrick-mead.'

1. The game requires a chess-board, five white and five black men, and something, such as a coin, with which to mark a square. The twenty-eight border-squares form 'the border': the other thirty-six form 'the field.'

2. The men are moved like chess-queens, that is, along any line of squares, straight or slanting.

3. The mark may be set on any square in the field. The marked square forms, with the surrounding eight, a 'rendezvous,' into which both Players try to get their men: it counts as a vacant square, so that a man may be moved into or over it.

4. When playing for a rendezvous, he who plays first may not move in that turn more than two squares: in any other turn a Player may not move more squares than he has men on the board. He may move all these squares with one man, or may divide them among two or more men. He may move a man

more than once in one turn, provided it be along the same line of squares.

5. He who begins sets on the border one of his men : the other does the same, and so on alternately. The Player named first in this Rule then sets the mark, taking care that none of his men are in the rendezvous and that he cannot, in one turn, move them all in. Both then play for this rendezvous, he who did not set the mark playing first.

6. As soon as either Player has 'won' the rendezvous, i.e. has all his men in it, he takes off the board some one of his adversary's men who is not in, and moves to the border, in accordance with Rule 2, all others of his adversary's men who are not in and are not already on the border.

7. Then, if the Players have unequal numbers of men in the field, he who has most makes the numbers equal by moving to the border one or more of his men.

8. Then he who has fewest men on the board, or in case of equality he who lost the last rendezvous, moves to the border one of his men, unless all be already on it : the other does the same, and so on alternately. The Player named first in this Rule then sets the mark as in Rule 5, and the game proceeds as before till one Player has taken four of his adversary's men, which wins the game.

October, 1881.

⌒ Mischmasch ⌒

Mischmasch is a game based on a form of word play which in recent years has occupied the minds of many contributors to the magazine *Word Ways*, edited by Ross Eckler of Spring Valley Road, Morristown, New Jersey 07960. One is given a doublet, triplet, or quartet of letters, then tries to discover a common English word in which those letters appear adjacent to one another. Carroll took the game's title from the title of one of the little magazines he wrote and printed irregularly when he was a youth—it is a German word equivalent to "hodgepodge."

Carroll first published the game in *The Monthly Packet* (June 1881) and later gave a revised version in the same magazine (November 1882). His anonymous three-page pamphlet, *Mischmasch: A Word-Game For Two Players Or Two Sets of Players,* was printed in 1882. It is a revised reprint of his November 1892 article in *The Monthly Packet.*

MISCHMASCH.

A WORD-GAME FOR TWO PLAYERS OR TWO SETS OF PLAYERS.

'Pars pro toto.'

THE essence of this game consists in one Player proposing a 'nucleus' (*i.e.* a set of two or more letters, such as 'gp,' 'emo,' 'imse'), and in the other trying to find a 'lawful word' (*i.e.* a word known in ordinary society, and not a proper name), containing it. Thus, 'magpie,' 'lemon,' 'himself,' are lawful words containing the nuclei 'gp,' 'emo,' 'imse.'

A nucleus must not contain a hyphen (*e.g.* for the nucleus 'erga,' 'flower-garden' is not a lawful word).

Any word, that is always printed with a capital initial (*e.g.* 'English'), counts as a proper name.

RULES.

1. Each thinks of a nucleus, and says 'ready' when he has done so. When both have spoken, the nuclei are named. A Player may set a nucleus without knowing of any word containing it.

2. When a Player has guessed a word containing the nucleus set to him (which need not be the word

thought of by the Player who set it), or has made up his mind that there is no such word, he says 'ready,' or 'no word,' as the case may be: when he has decided to give up trying, he says 'I resign.' The other must then, within a stated time (*e.g.* 2 minutes), say 'ready,' or 'no word,' or 'I resign,' or 'not ready.' If he says nothing, he is assumed to be 'not ready.'

3. When both have spoken, if the first speaker said 'ready,' he now names the word he has guessed: if he said 'no word,' he, who set the nucleus, names, if he can, a word containing it. The other Player then proceeds in the same way.

4. The Players then score as follows :—(N.B.—When a Player is said to 'lose' marks, it means that the other scores them.)

> Guessing a word, rightly, scores 1.
>   „   „   wrongly, loses 1.
> Guessing 'no word,' rightly, scores 2.
>   „   „   wrongly, loses 2.
> Resigning   loses 1.

This ends the first move.

5. For every other move, the Players proceed as for the first move, except that when a Player is 'not ready,' or has guessed a word wrongly, he has not a new nucleus set to him, but goes on guessing the one already in hand, having first, if necessary, set a new nucleus for the other Player.

6. A 'resigned' nucleus cannot be set again during the same game. If, however, one or more letters be added or subtracted, it counts as a new one.

7. The move, in which either scores 10, is the final one; when it is completed, the game is over, and the highest score wins, or, if the scores be equal, the game is drawn.

November, 1882.

[143]

⌒Syzygies⌒

"Invented a new way of working one word into another." Carroll wrote in his diary (December 12, 1879). "I think of calling the puzzle 'Syzygies.'" More than a decade elapsed before he records (April 14, 1891) sending an account of the game to *Vanity Fair*. He adds, "My first idea of this new puzzle occurred in 1879, but the scoring was too complex."

Evidently *Vanity Fair* did not accept the contribution. It appeared in *The Lady* (July 23, 1891), followed in later issues by letters and reports of Syzygy tournaments. In 1891 Carroll reprinted the article, with additions, as a four-page pamphlet bearing his name. His final and most lengthy account of the game was included in his 16-page booklet *Syzygies and Lanrick* (1893), a work reprinted in *The Lewis Carroll Picture Book*.

Syzygies is an elaboration of doublets. If two words contain inside them the same subset of consecutive letters, the set is called a "syzygy." In his 1893 booklet Carroll illustrated this by showing how WALRUS and SWALLOW are "yoked" together by the syzygy WAL. The idea is to link two associated words (which need not be the same length, as in doublets) in a chain such that every pair of adjacent words is joined by a syzygy. Carroll shows how WALRUS and CARPENTER can be linked as follows: WALRUS, PERUSE, HARPER, CARPENTER. A method of scoring chains is provided.

Carroll taught the game to many child-friends, and he mentions it often in his letters. In an 1891 letter to Beatrice Earle, he linked her first and last name by this chain: BEATRICE, THEATRICALS, MEDICAL, HANDICAPPED, APPEAR, PEARL, EARLE.

SYZYGIES.

—o—

A WORD-PUZZLE.

By Lewis Carroll.

[*Reprinted from* "The Lady" *for July 23rd, 1891.*]

WHEN two words have one or more letters standing together in the same order, common to both, this collection of letters may be called a "Syzygy" between the two words. Thus "a" is a Syzygy between cat and rat; "en" is a Syzygy between friend and enemy; and "din" is a Syzygy between pudding and dinner.

The puzzle consists in linking together two given words by a chain of words, called links, such that every two consecutive words may contain a Syzygy, and the longer the Syzygies are, the more marks do they obtain. Thus, supposing the two given words were door and window, the following would be a chain linking them together:—

DOOR
(oor)
poorest
(res)
resound
(und)
undo
(ndo)
WINDOW

The above will serve as a specimen of the way
in which such chains should be written out, each
Syzygy being placed in a parenthesis.

Rules.

1. A Syzygy may stand at the beginning, or end,
or in the middle of a word; but it may not begin
both of the words to which it belongs, neither may
it end both. Thus

> handsome
> (some)
> somewhere

would be a lawful Syzygy; but

> handsome
> (some)
> troublesome

would be an unlawful one. A chain containing an
unlawful Syzygy would get no marks.

2. The words used as links must be ordinary words
given in dictionaries, or inflexions of them. Proper
names and words containing hyphens are not allowed.

3. The letters "y" and "i" are to be regarded
as the same. Thus

> busy
> (usi)
> using

would be a lawful Syzygy.

4. The marks to be given with each chain are
calculated by adding together the number of letters
in the longest Syzygy and seven times the number
in the shortest, and deducting a mark for every link
and for every "waste" letter (*i.e.*, every letter which
does not enter into a Syzygy).

SPECIMEN CHAINS.

In the following chains the figure placed against each word indicates the number of "waste" letters in it.

CONVEY 4   *Score.*
(on)   2 + 14 16
Once ....o   1 + 8 9
(ce)   —
PARCEL.........4   7

———

CONVEY.........3
(nve)
Inverse 1   *Score.*
(rse)   3 + 21 24
Sparsely...........3   2 + 10 12
(par)   —
PARCEL.........3   12

———

CONVEY.........1
(conve)
Unconverted 5
(conver)   *Score.*
Converse..........1   6 + 28 34
(vers)   4 + 11 15
Versifier...........1   —
(fier)   19
Fiercely...........1
(rcel)
PARCEL.........2

———

We purpose to give a prize of One Guinea to the competitor who gains the greatest number of marks from the date of this number to the 29th of September. The first Competition will be to

Change a CONSERVATIVE *into a* LIBERAL.

———

Answers must be received by (or before) the first post on the 4th of August. The envelope to be marked "Syzygies" in the left-hand corner.

☞ Circular Billiards ☜

Carroll thought of playing billiards on a circular table in 1889, and first published its rules the following year as a single sheet printed on both sides and bearing his name. A second printing (the one given here) differed from the first only in extending the table for scoring. A third issue also differs slightly from the two previous ones.

Roger Green, in *The Lewis Carroll Handbook,* says he was told by Miss Menella Dodgson that a circular billiard table was actually made for Carroll. The *New York Times* (July 1, 1964) ran a full-page ad for Elliptipool, played on an elliptical table with a single pocket at one of the two foci. The ad said that on the following day the game would be demonstrated at Stern's department store by movie stars Paul Newman and his wife Joanne Woodward.

The eleventh edition of *Encyclopaedia Britannica,* in an article on billiards, reported that in 1907 an oval table (without pockets) was introduced in England. In 1964 a design patent (No. 198,571) was issued to Edwin E. Robinson for a circular pool table with four pockets.

CIRCULAR BILLIARDS,

FOR TWO PLAYERS.

INVENTED, IN 1889, BY

LEWIS CARROLL.

___

The Table is circular, with a cushion all round it, no pockets, and three white spots arranged in an equilateral triangle.

___

Rules.

1.

String for lead. Then the player, who is not to begin, places the 3 balls (red, white, and spot-white) on the spots provided for them. The first stroke must be played at the *red* ball.

2.

A 'miss' counts 1 to the adversary. A ball driven off the table, counts 2 to the adversary, and must be replaced on its original spot.

3.

If the ball in play strike one ball, and nothing else, it counts nothing.

4.

A cannon counts 2, and gives the right of playing again.

5.

Striking the cushion counts 1 for every ball struck afterwards. Thus, a cushion struck before striking one ball counts 1 : a cushion struck during a cannon counts 1 : a cushion struck previous to a cannon counts 2. Two or more consecutive cushions are reckoned as one only.

6.

Game is 50 or 100.

[P. T. O.

Remarks.

The circular Table will be found to yield an interesting variety of Billiard-playing, as the rebounds from the cushion are totally different from those of the ordinary game.

The 5 possible modes of scoring are here appended. (N. B. '*B*' stands for 'Ball', '*c*' for 'cushion'.)

All scores below the line give the right of playing again.

c B	scores	1
B B	,,	2
B c B	,,	3
c B B	,,	4
c B c B	,,	5

Selected Bibliography

Abeles, Francine. 1994. *The Mathematical Pamphlets of Charles Lutwidge Dodgson and Related Pieces.* New York: The Lewis Carroll Society of North America.

Bartley III, William W. 1977. *Lewis Carroll's Symbolic Logic.* New York: Clarkson Potter.

Carroll, Lewis. 1958. *Pillow Problems and a Tangled Tale.* New York: Dover.

Carroll, Lewis. 1958. *Symbolic Logic and the Game of Logic.* New York: Dover.

Carroll, Lewis. 1960. *The Humorous Verse of Lewis Carroll.* New York: Dover.

Carroll, Lewis. 1965. *The Works of Lewis Carroll.* Roger Lancelyn Green (ed.). London: Paul Hamlyn.

Carroll, Lewis. 1971. *The Rectory Umbrella and Mischmasch.* New York: Dover.

Carroll, Lewis. 1982. *The Complete Illustrated Works of Lewis Carroll*. Edward Guiliano (ed.). New York: Avenel.

Carroll, Lewis. 1988. *Sylvie and Bruno*. New York: Dover.

Clark, Anne. 1979. *Lewis Carroll: A Biography*. New York: Schocken.

Clark, Anne. 1981. *The Real Alice*. New York: Stein and Day.

Cohen, Morton (ed.). 1979. *The Letters of Lewis Carroll*. London: Oxford University Press.

Cohen, Morton. 1989. *Lewis Carroll: Interviews and Recollections*. Iowa City: University of Iowa Press.

Cohen, Morton. 1995. *Lewis Carroll: A Biography*. New York: Knopf.

Collingwood, Stuart Dodgson. 1898. *The Life and Letters of Lewis Carroll*. New York: Century.

Collingwood, Stuart Dodgson. 1961. *The Lewis Carroll Picture Book*. New York: Dover.

Fisher, John. 1973. *The Magic of Lewis Carroll*. New York: Simon and Schuster.

Gardner, Martin. 1960. *The Annotated Alice*. New York: Clarkson Potter; 1993. Revised edition, 1993. New York: Wings.

Gardner, Martin. 1990. *More Annotated Alice*. New York: Random House.

Gordon, Colin. 1982. *Beyond the Looking Glass*. New York: Harcourt Brace Jovanovich.

Green, Roger Lancelyn (ed.). 1954. *The Diaries of Lewis Carroll*. London: Oxford University Press.

Guiliano, Edward (ed.). 1976. *Lewis Carroll Observed*. New York: Clarkson Potter.

Guiliano, Edward (ed.). 1982. *Lewis Carroll: A Celebration*. New York: Clarkson Potter.

Guiliano, Edward and James Kincaid. 1982. *Soaring with the Dodo*. New York: The Lewis Carroll Society of North America.

Guiliano, Edward. 1986. *Lewis Carroll: An Annotated Bibliography, 1960–1977*. Charlottesville: University of Virginia Press.

Heath, Peter. 1974. *The Philosopher's Alice*. New York: St. Martin's Press.

Hudson, Derek. 1977. *Lewis Carroll: An Illustrated Biography.* New York: Meridian.

Huxley, Francis. 1976. *The Raven and the Writing Desk.* New York: Harper and Row.

Phillips, Robert (ed.). 1971. *Aspects of Alice.* New York: Vanguard.

Wakeling, Edward. 1992. *Lewis Carroll's Games and Puzzles.* New York: Dover.

Wakeling, Edward (ed.). 1993. *The Oxford Pamphlets, Leaflets, and Circulars of Charles Lutwidge Dodgson.* Charlottesville: University of Virginia Press.

Wakeling, Edward (ed.). 1993, 1994, 1995. *Lewis Carroll's Diaries, Vols. 1, 2, and 3.* Luton, England: Lewis Carroll Society.

Williams, Sidney and Falconer Madan. 1962. *The Lewis Carroll Handbook,* revised by Roger Lancelyn Green. London: Oxford University Press.

Index

CPSIA information can be obtained at www.ICGtesting.com
Printed in the USA
LVOW10s2245290415

436598LV00005B/85/P